Texts and Monographs in Physics

Springer
Berlin
Heidelberg
New York
Barcelona
Hong Kong
London
Milan
Paris
Singapore
Tokyo

Texts and Monographs in Physics

Series Editors: R. Balian W. Beiglböck H. Grosse E. H. Lieb N. Reshetikhin H. Spohn W. Thirring

From Microphysics to Macrophysics I + II Methods and Applications of Statistical Physics By R. Balian

Variational Methods in Mathematical Physics A Unified Approach By P. Blanchard and E. Brüning

Multi-Hamiltonian Theory of Dynamical Systems By M. Błaszak

Quantum Mechanics: Foundations and Applications 3rd enlarged edition By A. Böhm

The Early Universe Facts and Fiction 3rd corrected and enlarged edition By G. Börner

Operator Algebras and Quantum Statistical Mechanics I + II 2nd edition By O. Bratteli and D. W. Robinson

Statistical Methods in Quantum Optics 1. Master Equations and Fokker–Planck Equations By H. J. Carmichael

Geometry of the Standard Model of Elementary Particles By. A. Derdzinski

Scattering Theory of Classical and Quantum *N*-Particle Systems By. J. Derezinski and C. Gérard

Effective Lagrangians for the Standard Model By A. Dobado, A. Gómez-Nicola, A. L. Maroto and J. R. Peláez

Quantum The Quantum Theory of Particles, Fields, and Cosmology By E. Elbaz

Quantum Relativity A Synthesis of the Ideas of Einstein and Heisenberg By D. R. Finkelstein

Quantum Mechanics I + II By A. Galindo and P. Pascual

The Elements of Mechanics By G. Gallavotti

Local Quantum Physics Fields, Particles, Algebras 2nd revised and enlarged edition By R. Haag

***CP* Violation Without Strangeness** Electric Dipole Moments of Particles, Atoms, and Molecules By I. B. Khriplovich and S. K. Lamoreaux

Quantum Groups and Their Representations By A. Klimyk and K. Schmüdgen

Inverse Schrödinger Scattering in Three Dimensions By R. G. Newton

Scattering Theory of Waves and Particles 2nd edition By R. G. Newton

Quantum Entropy and Its Use By M. Ohya and D. Petz

Generalized Coherent States and Their Applications By A. Perelomov

Essential Relativity Special, General, and Cosmological Revised 2nd edition By W. Rindler

Path Integral Approach to Quantum Physics An Introduction 2nd printing By G. Roepstorff

Renormalization An Introduction By M. Salmhofer

Finite Quantum Electrodynamics The Causal Approach 2nd edition By G. Scharf

From Electrostatics to Optics A Concise Electrodynamics Course By G. Scharf

The Mechanics and Thermodynamics of Continuous Media By M. Šilhavý

Large Scale Dynamics of Interacting Particles By H. Spohn

The Theory of Quark and Gluon Interactions 3rd revised and enlarged edition By F. J. Ynduráin

Relativistic Quantum Mechanics and Introduction to Field Theory By F. J. Ynduráin

Quantum Field Theory in Condensed Matter Physics By N. Nagaosa

Naoto Nagaosa

Quantum Field Theory in Strongly Correlated Electronic Systems

Translated by Stefan Heusler

With 18 Figures

Springer

Professor Naoto Nagaosa
Department of Applied Physics
The University of Tokyo
Bunkyo-ku, Tokyo 113, Japan

Translator:
Stefan Heusler
Bothestrasse 106
D-69126 Heidelberg, Germany

Library of Congress Cataloging-in-Publication Data.
Nagaosa, N. (Naoto), 1958- [Denshi sōkan ni okeru ba no ryōshiron. English] Quantum field theory in strongly correlated electronic systems / Naoto Nagaosa ; translated by Stefan Heusler. p. cm. – (Texts and monographs in physics) Includes bibliographical references and index. ISBN 3-540-65981-1 (alk. pager) 1. Quantum field theory. 2. Superconductivity. I. Title. II. Series. QC174.45.N2513 1999 530.14'3–dc21 99-16714

Title of the original Japanese edition:
Quantum Field Theory in Strongly Correlated Electronic Systems, Vol. II by Naoto Nagaosa

Originally published in Japanese by Iwanami Shoten, Publishers, Tokyo in 1998

ISSN 0172-5998
ISBN 3-540-65981-1 Springer-Verlag Berlin Heidelberg New York

Typesetting: Data conversion by EDV-Beratung F. Herweg, Hirschberg
Cover design: *design & production* GmbH, Heidelberg
SPIN: 10701991 55/3144/di - 5 4 3 2 1 0 – Printed on acid-free paper

Preface

Research on electronic systems in condensed matter physics is at present developing very rapidly, where the main focus is changing from the "single-particle problem" to the "many-particle problem". That is, the main research interest changed from phenomena that can be understood in the single-particle picture, as, for example, in band theory, to phenomena that arise owing to the interaction between many electrons.

As examples of the latter case, we mention superconductivity and magnetism; in both cases the research has a long history. New developments in these fields are the studies on phenomena that are beyond the scope of mean field theories such as BCS theory and mean field theory of the spin density wave – and are related to research on so-called electronic correlation. Electronic correlation effects arise owing to strong quantum as well as thermal fluctuations. When fluctuations are large, the interaction between different degrees of freedom becomes important; for example, the interplay between magnetism and superconductivity in high temperature superconductors.

The best framework to describe strongly interacting degrees of freedom – which is nothing but the "field" itself – is quantum field theory. In this volume, applications of quantum field theory to the problem of strongly correlated electronic systems are presented in a – hopefully – systematic way in order to be understandable to the beginner. Knowledge of the basic topics discussed in *Quantum Field Theory in Condensed Matter Physics*, written by the same author, is presumed.

This volume consists of a series of themes. In the first part, one-dimensional, many-particle quantum systems are discussed. In a single dimension, order cannot emerge owing to strong quantum fluctuations, and therefore, down to zero temperature, a quantum liquid survives. Therefore, the one-dimensional system is a very important toy model for the discussion of electronic correlation, where many ideas and methods can successfully be applied, and where the results are established with the highest accuracy. Discussion of these models determines the basic direction for the whole range of problems related to electronic correlation. In Chap. 1, the XXZ quantum spin chain is discussed in its classical and quantum limits. The most important object here is the kink, which can be described in terms of a fermion that is related to the spins by a non-local phase factor (a Jordan–Wigner transformation).

Here, issues such as a new particle (kink), non-locality and quantum statistics appear for the first time, being important in all problems related to electronic correlation.

In Chap. 2, the quantum field theory of an interacting fermion system, being equivalent to the XXZ spin chain, is discussed. Starting from the fact that the one-dimensional fermion system can be described by the two Fermi points $k_{\mathrm{F}}^{\mathrm{R}}$ and $k_{\mathrm{F}}^{\mathrm{L}}$, the canonical conjugate relation between the density and the current is derived. Using the (bosonic) phase fields θ_{+} and θ_{-} for their description, the fermionic system is finally mapped to the Sine–Gordon system. Kinks (solitons) are the classical solution of the quantum Sine–Gordon system, which correspond to the original fermions. The kink connects the different minima of the sine potential, which has a finite excitation energy in a classical approximation. However, when the quantum fluctuations become large, the sine potential itself is effectively reduced and eventually washed out, and in the excitation spectrum of the kink, the gap disappears. This will be demonstrated in Sect. 2.1 using the renormalization group. When the sine potential can effectively be neglected, the system can simply be described in terms a of Gaussian $(1+1)$-dimensional free bosonic theory. At first sight, the theory looks trivial; however, it contains very fundamental theoretical structures, that is, invariance under conformal (angle conserving) transformations on the complex plane. In Sect. 2.2, the Gaussian theory is discussed using conformal field theory.

In this way, many different theoretical approaches can be used to describe gapless one-dimensional quantum liquids. However, quantum liquids having a gap in the excitation spectrum are also known. The antiferromagnetic Heisenberg model with integer spin S, called the Haldene system, is a representative example. For its analysis we use another method different from bosonization, based on the non-linear sigma model (Sect. 2.3). The advantages of this model are that the meaning of the Berry phase can be seen very clearly, and that it can be generalized to higher dimensions.

Starting from Chap. 3, systems containing both a charge and spin degree of freedom, and higher-dimensional systems are discussed. First, strongly correlated electronic systems are introduced. In Sect. 3.1, many different models are presented, and the idea of deriving effective Hamiltonians by restricting the Hilbert space is introduced. Because these restrictions in the Hilbert space are represented by constraint conditions, strongly correlated electronic systems are often formulated as quantum theory under constraints.

Spin-charge separation is one of the central issues in strongly correlated electronic systems. This phenomenon appears in the one-dimensional interacting electron system in the most striking way, as described in Sect. 3.2 using a generalization of the bosonization method introduced in Chap. 2. Density and current of spin $\uparrow$ and $\downarrow$ are defined, where the sum and the difference represent the charge and spin degree of freedom, respectively. This spin charge separated one-dimensional quantum liquid – the Tomonaga–Luttinger fluid –

is the representative example of a non-Fermi liquid. Here, the description using bosonization is not principally different from that of the theory of a Fermi liquid; however, because the Fermi surface in a single dimension only consists of two single points, no individual excitation exists, but only collective excitation modes. This is the reason for the non-Fermi-liquid behaviour.

On the other hand, in two and three dimensions, compared with a single dimension, the quantum fluctuations are smaller, and under some circumstances magnetic ordering occurs. For this reason, one crucial degree of freedom for the description of strongly correlated electronic systems is the magnetic moment, and it seems natural to describe the properties of the system by analysing its ordering and fluctuations. Following this outline, in Sects. 3.3 and 3.4, mean field theory and the fluctuations of the magnetic ordering are discussed, respectively. In particular the latter is analysed in terms of both self-consistent renormalization and the quantum renormalization group in order to investigate the singular enhancement of quantum fluctuations in the vicinity of phase transition points at zero temperature – driven by some external parameters like pressure, etc.

Another point of view beside the discussion of magnetic moment and its fluctuations is the discussion of spin singlet formation. One representative example is the magnetic impurities in the metal – the Kondo effect. When only one spin moment exists, magnetic ordering cannot occur. Finally, the localized spin and the spins of conduction electrons of the metal form a singlet and the entropy is quenched. As a result, the system becomes a magnetically inert local Fermi liquid. This state can be analysed in terms of the slave boson method. On the other hand, in the case when the number of channels of conduction electrons is large (the multi-channel Kondo problem), it is known that a local non-Fermi-liquid state can arise also. This state, too, can be described using the slave boson method. By determining a saddle point solution where the bosons do not condensate, it can be shown that the Green's function is characterized by non-trivial critical exponents.

The Kondo model can also be approached using many other theoretical methods. In particular, by partial wave analysis, the incoming wave and the scattered wave can be regarded as one-dimensional in dependence on the polar coordinate r. In such a way, the problem of a localized impurity can be mapped on the problem of an impurity interacting with an electron in one dimension. Here, the framework of the $(1+1)$-dimensional quantum field theory as developed in Chap. 2 can be applied. Furthermore, it can be shown that the impurity problem is equivalent to the effective model developed in Sect. 4.2 using dynamic mean field theory in the limit of large spatial dimension d. With this knowledge, it becomes clear that the problem of electron correlation in $d = 0$ dimensions (local impurity), $d = 1$ and $d = \infty$ are intimately related to each other.

However, this does not mean that all systems with strongly correlating electrons are understood. In particular, it is possible that in the important dimensions $d = 2$ and $d = 3$ physical phenomena could appear that cannot be explained within this framework. In the theories described above, in every case the degree of freedom on the site – for example, the localized moment or the localized electrons appearing in dynamic mean field theory – has been considered. One might assume that it is a complementary approach to focus on the degree of freedom defined on the link between two sites. The field defined on the link describes the "relation" between the degrees of freedom on each site; mathematically speaking, this is the connection and, in physics, this field is called the gauge field. For example, when the spin moment is considered as the degree of freedom at each site, for a quantum liquid where, owing to quantum fluctuations, the magnetic order has disappeared, we might assume rather that the singlet amplitude defined on the link is a meaningful order parameter. In Chap. 5, theories of strongly correlated electronic systems are developed from this point of view. Explicitly, the quantum anti-ferromagnet (Sect. 5.1), the doped Mott insulator, being deeply related to high temperature superconductivity (Sect. 5.2), and the quantum Hall liquid (Sect. 5.3) are described in terms of gauge theory. In particular, in Sect. 5.3, the use of the Chern–Simons gauge field for the description of non-local quantum statistics is discussed, being the generalization of the Jordan–Wigner transformation of Chap. 1 to the two-dimensional case. In such a way, the book closes with a reprise, and the author hopes that the reader will be able to see the content of this book in a new light from the point of view of gauge theories.

I owe special thanks to my supervisors and colleagues, especially E. Hanamura, Y. Toyazawa, P. A. Lee, H. Fukuyama, S. Tanaka, M. Imada, K. Ueda, S. Uchida, Y. Tokura, N. Kawakami, A. Furusaki, T. K. Ng, and Y. Kuramoto.

Tokyo, January 1999 *Naoto Nagaosa*

Table of Contents

1. The One-Dimensional Quantum Spin Chain

Among many-body quantum systems, the one-dimensional occupies a special position. Owing to strong quantum fluctuations, the system is a quantum liquid, and down to zero temperature, no ordered state emerges. In this chapter, the one-dimensional quantum spin system will be discussed, being the introduction to the theory of strongly correlated systems.

1.1 The $S = 1/2$ *XXZ* Spin Chain

First, we consider the following Hamiltonian:

$$H = J_\perp \sum_i \left(S_i^x S_{i+1}^x + S_i^y S_{i+1}^y\right) + J_z \sum_i S_i^z S_{i+1}^z \,. \tag{1.1.1}$$

Here, $\boldsymbol{S}_i$ is the spin $S = 1/2$ operator. Expressing the $\uparrow$ spin state as $[1, 0]^t$ and the $\downarrow$ state as $[0, 1]^t$, and using the Pauli matrices $\boldsymbol{\sigma} = (\sigma^x, \sigma^y, \sigma^z)$

$$\sigma^x = \begin{bmatrix} 0 & 1 \\ 1 & 0 \end{bmatrix}, \qquad \sigma^y = \begin{bmatrix} 0 & -\mathrm{i} \\ \mathrm{i} & 0 \end{bmatrix}, \qquad \sigma^z = \begin{bmatrix} 1 & 0 \\ 0 & -1 \end{bmatrix},$$

the spin operator is given by $\boldsymbol{S}_i = (\hbar/2)\boldsymbol{\sigma}_i$. The spin operator satisfies

$$\left[S_i^\alpha, S_j^\beta\right] = \mathrm{i}\hbar\delta_{ij}\varepsilon_{\alpha\beta\gamma}S_i^\gamma \,. \tag{1.1.2}$$

$\varepsilon_{\alpha\beta\gamma}$ is the total antisymmetric tensor with $\varepsilon_{xyz} = 1$. In what follows, we usually set $\hbar = 1$. $J_\perp$ as well as J_z are the nearest-neighbour exchange interactions; and for the case $J_\perp = J_z$, the Hamiltonian is called the Heisenberg model.

From the commutation relation (1.1.2), it is clear that the unitary operator

$$U_i(\boldsymbol{n}, \theta) = \mathrm{e}^{\mathrm{i}\theta \boldsymbol{n}\cdot\boldsymbol{S}_i} \tag{1.1.3}$$

rotates the spin. Here, $\boldsymbol{n}$ is the unit vector, and θ is the rotation angle. For example, the case where $\boldsymbol{n} = \boldsymbol{e}_z = (0, 0, 1)$,

$$U_i(\boldsymbol{e}_z, \theta) S_i^\alpha U_i^\dagger(\boldsymbol{e}_z, \theta) = \tilde{S}_i^\alpha(\theta)\,, \tag{1.1.4}$$

leads to

$$\frac{d\tilde{S}_i^x(\theta)}{d\theta} = U_i \cdot i\,[S_i^z, S_i^x]\,U_i^\dagger = -\tilde{S}_i^y(\theta)\,, \tag{1.1.5a}$$

$$\frac{d\tilde{S}_i^y(\theta)}{d\theta} = \tilde{S}_i^x(\theta)\,, \tag{1.1.5b}$$

$$\frac{d\tilde{S}_i^z(\theta)}{d\theta} = 0\,. \tag{1.1.5c}$$

The solution is then given by

$$\tilde{S}_i^x(\theta) = S_i^x \cos\theta - S_i^y \sin\theta\,, \tag{1.1.6a}$$

$$\tilde{S}_i^y(\theta) = S_i^x \sin\theta + S_i^y \cos\theta\,, \tag{1.1.6b}$$

$$\tilde{S}_i^z(\theta) = S_i^z\,. \tag{1.1.6c}$$

Obviously, this describes a rotation around the z-axis. Next, constructing the unitary operator $T = \prod_{n:\mathrm{odd}} U_n(\boldsymbol{e}_z, \pi)$ from U_i, because of

$$TS_i^x T^\dagger = (-1)^i S_i^x\,, \tag{1.1.7a}$$

$$TS_i^y T^\dagger = (-1)^i S_i^y\,, \tag{1.1.7b}$$

$$TS_i^z T^\dagger = S_i^z\,. \tag{1.1.7c}$$

the transformation $THT^\dagger$ of (1.1.1) interchanges $J_\perp \to -J_\perp, J_z \to J_z$. We conclude that the sign of $J_\perp$ is not essential.

On the other hand, the sign of J_z plays an essential role for the quantum system. This is owing to the fact that, different from (1.1.7), the commutation relation (1.1.2) of the spin components is changed for $\boldsymbol{S}_i \to -\boldsymbol{S}_i$. First, we construct with S^x and S^y:

$$S^\pm = S^x \pm iS^y\,. \tag{1.1.8}$$

Because of

$$S^+ = \begin{bmatrix} 0 & 1 \\ 0 & 0 \end{bmatrix}, \qquad S^- = \begin{bmatrix} 0 & 0 \\ 1 & 0 \end{bmatrix}.$$

S^+ is the operator that flips the spin ↑ state to ↓, and S^- flips ↓ to ↑. Using these operators, the Hamiltonian (1.1.1) can be re-expressed as

$$H = \frac{J_\perp}{2} \sum_i (S_i^+ S_{i+1}^- + S_i^- S_{i+1}^+) + J_z \sum_i S_i^z S_{i+1}^z\,. \tag{1.1.9}$$

Now, regarding S_i^z as a 'coordinate', $S_i^\pm$ displace this coordinate as if it were the 'momentum'. In this interpretation, the unitary operator U_i (1.1.3) corresponds to the linear transformation operator $U = e^{i\alpha p}$ (x: coordinate; p: momentum; α: constant). Then, the term proportional to $J_\perp$ in (1.1.9) represents the 'kinetic energy' causing the quantum fluctuations of S_i^z, and the term proportional to J_z represents the 'potential energy' that causes the ordering of S_i^z. The competition between these two tendencies is the physics that is contained in (1.1.1) as well as in (1.1.9).

We first consider the classical limit $J_\perp = 0$. This is the so-called Ising-model; and the spins align at zero temperature depending on the sign of J_z ferromagnetic ally ($J_z < 0$), or anti-ferromagnetically ($J_z > 0$). Notice that the ground state is two-fold degenerate because the Hamiltonian is invariant under the transformation $S_z^i \to -S_z^i$, performed at all sites i. Calling these two ground states A and B and assuming that the system at the right-hand side is in state A, and at the left-hand side in state B, then somewhere there must exist a boundary between region A and region B. This boundary is called a kink or soliton. Because at finite temperature this excitation occurs with a finite density, the spin correlation function $F(r) = \langle S_i^z S_{i+r}^z \rangle$ will decay exponentially with a correlation length ξ.

Let us determine this explicitly. We align N spins from $i = 1$ to $i = N$ with free ends. In the thermodynamic limit $N \to \infty$, we can ignore the influence of the boundary. We first assume that the spin S_1^z at the left side is fixed to, say, $1/2$. Then, instead of S_2^z, it is possible to consider $L_{12} \equiv S_1^z S_2^z$ as variable. In the same manner, instead of S_3^z, defining $L_{23} \equiv S_2^z S_3^z$ and so on, the statistical mechanics can be formulated in terms of $L_{i,i+1}$ instead of S_i^z. The correlation between S_i^z and S_{i+1}^z, being defined on the site, is expressed by $L_{i,i+1}$, being defined on the link. In the parallel or anti-parallel case, the obtained value is $\pm 1/4$, respectively. Now, we write $L_{i+1/2}$ instead of $L_{i,i+1}$. Then, the Ising model can be expressed as

$$H = J_z \sum_{i=1}^{N-1} L_{i+1/2} , \tag{1.1.10}$$

which is the independent sum of the energy at every link. The free energy of every site can be calculated easily:

$$\begin{aligned} f &= -\frac{1}{\beta} \lim_{N\to\infty} \frac{1}{N} \ln Z = -\frac{1}{\beta} \ln \sum_{L=\pm 1/4} \mathrm{e}^{-\beta J_z L} \\ &= -\frac{1}{\beta} \ln \left(2 \cosh \frac{\beta J_z}{4} \right) . \end{aligned} \tag{1.1.11}$$

Next, based on the above discussion, we calculate $F(r) = \langle S_i^z S_{i+r}^z \rangle$. First, we observe that $S_i^z S_{i+r}^z$ is 'non-local' when expressed in terms of $L_{j+1/2}$. That is,

$$4 S_i^z S_{i+r}^z = (4L_{i+1/2})(4L_{i+3/2}) \dots (4L_{i+r-1/2}) \tag{1.1.12}$$

can be expressed as the product of r terms in L connecting the sites i and $i + r$. With (1.1.10), from this fact we obtain

$$4F(r) = \prod_{j=1}^{r} \left[\frac{\sum\limits_{L=\pm 1/4} 4L\, \mathrm{e}^{-\beta J_z L}}{\sum\limits_{L=\pm 1/4} \mathrm{e}^{-\beta J_z L}} \right]$$
$$= \left[\tanh\left(\frac{-\beta J_z}{4} \right) \right]^r$$
$$= (-\operatorname{sgn} J_z)^r \exp\left[r \ln \tanh\left(\frac{\beta |J_z|}{4} \right) \right] . \quad (1.1.13)$$

Writing the absolute value of the left-hand side of (1.1.13) as $\mathrm{e}^{-r/\xi}$, the correlation length ξ is determined to be

$$\xi = -\frac{1}{\ln \tanh\left(\frac{\beta |J_z|}{4} \right)} . \quad (1.1.14)$$

Furthermore, at low temperature $\beta |J_z| \gg 1$, in the limit where a small number of kinks are thermally excited, equation (1.1.14) becomes

$$\xi^{-1} = 2\, \mathrm{e}^{-\beta |J_z|/2} . \quad (1.1.15)$$

Recalling that the creation energy for a kink is given by $\Delta E = |J_z|/2$, the right-hand side corresponds to the number of kinks, and ξ can be interpreted as the mean distance between two kinks. That is, the long-range order $F(r) = \frac{1}{4}(-\operatorname{sgn} J_z)^r$ at $T = 0\,\mathrm{K}$ is destroyed owing to kink excitation. Notice that a kink comes up as a change in sign of $L_{i+1/2}$.

The classical statistical mechanics described above almost did not depend on the sign of J_z. The only dependence emerges in $(-\operatorname{sgn} J_z)^r$ in (1.1.13), and, indeed, J_z changes to $-J_z$ when in the Hamiltonian all S_i^z are altered to $-S_i^z$. When proceeding to quantum mechanics, this is no longer correct. This is owing to the following reason. The term proportional to $J_\perp$ in (1.1.9) causing the quantum mechanical motion expresses the simultaneous flip of two neighbouring spins. It can be expressed as a product of S^+ and S^-, therefore when one S^z is growing, the other must diminish, and the sum $S_i^z + S_{i+1}^z$ is conserved. This simultaneous spin flip is the change between $|S_i^z = 1/2, S_{i+1}^z = -1/2\rangle$ and $|S_i^z = -1/2, S_{i+1}^z = 1/2\rangle$. For parallel spins, this change cannot occur. In a more mathematical language, let us define the total spin operator:

$$\boldsymbol{S}_{\mathrm{tot}} = \sum_i \boldsymbol{S}_i . \quad (1.1.16)$$

Then, $S_{\mathrm{tot}}^2 = \boldsymbol{S}_{\mathrm{tot}} \cdot \boldsymbol{S}_{\mathrm{tot}}$, and S_{tot}^z commute with the Hamiltonian (1.1.9):

$$[S_{\mathrm{tot}}^2, H] = [S_{\mathrm{tot}}^z, H] = 0 . \quad (1.1.17)$$

That is, the quantum system can be decomposed into eigenstates of S_{tot}^2 and S_{tot}^z. In the case when $J_z < 0$ of the ferromagnetic interaction, it is clear that the classical ground states $|\{S_i^z = 1/2\}\rangle$ and $|\{S_i^z = -1/2\}\rangle$ are also the eigenstates of the Hamilton operator H. On the other hand, the classical ground state in the anti-ferromagnetic case (the so-called Néel state) is not an eigenstate of the Hamiltonian. This becomes clear by constructing the staggered magnetization

$$\boldsymbol{S}_{\text{staggered}} = \sum_i (-1)^i \boldsymbol{S}_i \,. \tag{1.1.18}$$

The Néel state is an eigenstate of $S^z_{\text{staggered}}$. However, $[S^z_{\text{staggered}}, H]$ is different from zero. That is, for $S^z_{\text{staggered}}$ there must exist a zero point quantum fluctuation. The physical picture for this zero point fluctuation is just the resonance between the states $|S_i^z = 1/2, S_{i+1}^z = -1/2\rangle$ and $|S_i^z = -1/2, S_{i+1}^z = 1/2\rangle$ that has been mentioned before. Considering for simplicity only two spins, $\boldsymbol{S}_i$ and $\boldsymbol{S}_{i+1}$, setting $J_\perp > 0$, owing to this resonance the ground state becomes a linear combination of two states

$$|\text{singlet}\rangle = \frac{1}{\sqrt{2}}\left(\left|\frac{1}{2}, -\frac{1}{2}\right\rangle - \left|-\frac{1}{2}, \frac{1}{2}\right\rangle\right) . \tag{1.1.19}$$

This is the spin singlet wave function. That is, the quantum fluctuation of S_i^z leads to the singlet formation. Equation (1.1.9) can be interpreted as competition between the tendency of $J_\perp$ to create singlet states leading to the emergence of a quantum liquid, and the tendency of J_z to order the spins.

Above, considerations concerning the ground state have been done. Next, we consider the first excited state based on the Néel ground state of the classical system. This is the domain wall as shown in Fig. 1.1. In what follows, we think about the Ising limit $J_z \gg J_\perp > 0$, where mixing between states with different numbers of domain walls can be ignored. First, we consider the state containing one domain wall. Calling Ψ_n the wave function of the state where a domain wall is present between site n and $n+1$, we obtain

$$(H - E_{\text{Néel}})\Psi_n = \frac{J_z}{2}\Psi_n + \frac{J_\perp}{2}(\Psi_{n+2} + \Psi_{n-2}) \,. \tag{1.1.20}$$

Constructing a plane wave state of a domain wall

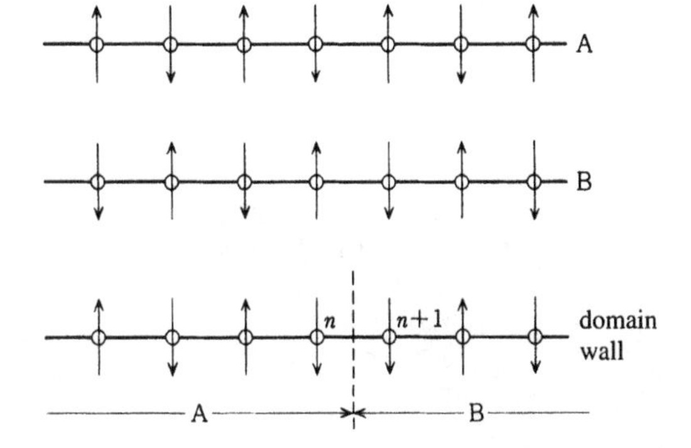

Fig. 1.1. The two degenerate classical Néel ground states A and B and the domain wall

$$\Psi(k) = \frac{1}{\sqrt{N}} \sum_n e^{ikn} \Psi_n , \tag{1.1.21}$$

we obtain

$$(H - E_{\text{Néel}})\Psi(k) = \left(\frac{J_z}{2} + J_\perp \cos 2k \right) \Psi(k) . \tag{1.1.22}$$

Therefore, the excitation energy is given by

$$\varepsilon_{\text{DW}}(k) = \frac{J_z}{2} + J_\perp \cos 2k . \tag{1.1.23}$$

When one domain wall is present (in general, an odd number), the sign of the staggered magnetization of the configuration at the boundaries $n = -\infty$ and $n = +\infty$ changes. Therefore, no long-range order is present in the Ising model at finite temperature as discussed above. Here, because we consider excited states at zero temperature, we set the staggered magnetization at $n = \pm\infty$ to be equal, corresponding to periodic boundary condition. Then, at least two domain walls have to be created. Calling the momenta k_1 and k_2, respectively, the total momentum q and the excitation energy ΔE are given by

$$\begin{aligned} q &= k_1 + k_2 \\ \Delta E &= \varepsilon_{\text{DW}}(k_1) + \varepsilon_{\text{DW}}(k_2) . \end{aligned} \tag{1.1.24}$$

The possible values for ΔE for every q are given by the oblique region in Fig. 1.2. In particular, the lower boundary is given by

$$\Delta E_{\text{LB}} = J_z - 2J_\perp |\cos q| . \tag{1.1.25}$$

We conclude that the excitation spectrum does not consist of isolated spin wave states, but is the continuum. reflecting the fact that the elementary excitations are kinks (domain walls).

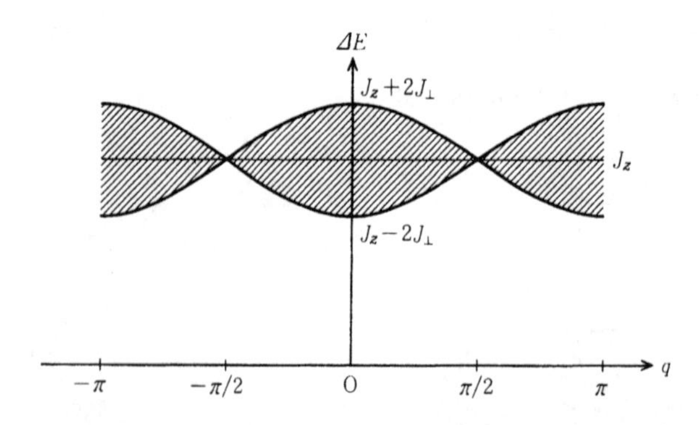

Fig. 1.2. The energy of the domain wall pair

1.2 The Jordan–Wigner Transformation and the Quantum Kink

In the previous section, the one-dimensional Ising model has been discussed as an example to demonstrate the importance of the kink. In this chapter, the quantum mechanics of the kink will be discussed in more detail. Starting with the conclusion, we note that the kink can be described in terms of fermions obtained by the following so-called Jordan–Wigner transformation

$$S_i^+ = S_i^x + \mathrm{i}S_i^y = f_i^\dagger K(i) = K(i) f_i^\dagger \,, \tag{1.2.1a}$$

$$S_i^- = S_i^x - \mathrm{i}S_i^y = K(i) f_i = f_i K(i) \,, \tag{1.2.1b}$$

$$S_i^z = f_i^\dagger f_i - \tfrac{1}{2} \,. \tag{1.2.1c}$$

Here, $K(i)$ is a non-local operator, defined by

$$K(i) = \exp\left[\mathrm{i}\pi \sum_{j=1}^{i-1} f_j^\dagger f_j\right] = \exp\left[\mathrm{i}\pi \sum_{j=1}^{i-1}\left(S_j^z + \frac{1}{2}\right)\right] \,. \tag{1.2.2}$$

$K(i) = K(i)^\dagger$, and $[K(i), K(j)] = 0$. Furthermore, for $i \leq j$, $K(i)$ and $\boldsymbol{S}_j$ also commute. From (1.2.1) we conclude:

$$\begin{aligned} f_i^\dagger &= S_i^+ K(i) \,, \\ f_i &= S_i^- K(i) \,. \end{aligned} \tag{1.2.3}$$

From these equations, the anti-commutation relation of fermions can be deduced.

We first examine the meaning of $K(i)$. Using the unitary operator U_i introduced in (1.1.3), we can write

$$K(i) = \left[\prod_{j=1}^{i-1} U_j(\boldsymbol{e}_z, \pi)\right] \mathrm{e}^{\frac{\mathrm{i}\pi}{2}(i-1)} \,. \tag{1.2.4}$$

We conclude that $K(i)$ rotates all spins from $j = 1$ to $i - 1$ by π around the z axis; that is, this operator shifts S^x and S^y to $-S^x$ and $-S^y$. $K(i)$ is a non-local spin rotating operator, and therefore creates a kink. Now, for $i < j$ we consider

$$f_i f_j = S_i^- K(i) S_j^- K(j) \,. \tag{1.2.5}$$

From the above considerations, we obtain

$$K^\dagger(j) S_i^- K(j) = -S_i^- \tag{1.2.6}$$

that is, $K(j)$ and S_i^- do anti-commute (for $j > i$). Finally, the right-hand side of (1.2.5) becomes

$$\begin{aligned} S_i^- K(i) S_j^- K(j) &= S_i^- S_j^- K(i) K(j) = S_j^- S_i^- K(i) K(j) \\ &= -S_j^- K(j) S_i^- K(i) = -f_j f_i \end{aligned} \tag{1.2.7}$$

and therefore

$$\{f_i, f_j\}_+ = 0 \tag{1.2.8}$$

holds. In the same manner, for $i < j$, because of

$$\begin{aligned} f_i f_j^\dagger + f_j^\dagger f_i &= S_i^- K(i) S_j^+ K(j) + S_j^+ K(j) S_i^- K(i) \\ &= S_i^- S_j^+ K(i) K(j) - S_j^+ S_i^- K(j) K(i) \\ &= \left[S_i^-, S_j^+\right] K(i) K(j) = 0 \, , \end{aligned}$$

we obtain

$$\{f_i, f_j^+\} = \delta_{ij} \, . \tag{1.2.9}$$

In this manner, the non-local spin rotating operator (in the xy-plane) translates the commutation relation of the spin operators at different sites into the anti-commutation relation of fermion operators. As follows from (1.2.1c), at one site, the two possible states $S_i^z = \pm 1/2$ correspond to fermion number $n_i = f_i^\dagger f_i = 1, 0$, respectively. We see how the Pauli principle naturally describes the limited possibilities for the allowed spin states. That these fermions correspond to kinks can be seen by considering, for example, the state $|F\rangle$ where all spins are in the state $S_i^x = +1/2$. Acting with $f_n^\dagger$, we obtain the state $f_n^\dagger |F\rangle$ with $S_i^x = -\frac{1}{2}$ $(1 \le i < n)$, $S_n^x = +\frac{1}{2}$, $S_i^z = +\frac{1}{2}$ $(i > n)$.

The Jordan–Wigner transformation works well because although (1.2.1) is non-local, the Hamiltonian (1.1.1) [(1.1.9)] can be expressed with these fermions in a local manner. First, we consider the term proportional to $J_\perp$:

$$\begin{aligned} S_i^+ S_{i+1}^- &= f_i^+ K(i) K(i+1) f_{i+1} \\ &= f_i^+ \exp\left[\mathrm{i}\pi f_i^\dagger f_i\right] f_{i+1} = f_i^+ f_{i+1} \, . \end{aligned} \tag{1.2.10}$$

Here, we used the fact that in the state multiplied with $\exp[\mathrm{i}\pi f_i^\dagger f_i]$, owing to $f_i^\dagger$, at the ith site no more fermion can be present. Also, the term proportional to J_z can be expressed locally using (1.2.1c), leading finally to

$$H = -\frac{J_\perp}{2} \sum_{i=1}^{N} \left(f_i^\dagger f_{i+1} + f_{i+1}^\dagger f_i\right) + J_z \sum_{i=1}^{N} \left(f_i^\dagger f_i - \frac{1}{2}\right) \left(f_{i+1}^\dagger f_{i+1} - \frac{1}{2}\right) . \tag{1.2.11}$$

As mentioned in Sect. 1.1, the sign of $J_\perp$ can be chosen freely, therefore we can put a minus sign before $J_\perp > 0$. We obtained the spinless fermion model with nearest site interaction J_z.

At this point, we discuss the periodic boundary conditions. Considering a ring and setting

$$\boldsymbol{S}_{N+1} = \boldsymbol{S}_1 \, , \tag{1.2.12}$$

the term $i = N$ proportional to $J_\perp$ in (1.2.11) becomes

$$\frac{J_\perp}{2}\left(S_N^+ S_1^- + S_N^- S_1^+\right) . \tag{1.2.13}$$

Here, we obtain

$$S_N^+ S_1^- = K(N) f_N^\dagger f_1 = -K f_N^\dagger f_1 . \tag{1.2.14}$$

The factor K can be expressed as

$$K = \exp\left[\mathrm{i}\pi \sum_{i=1}^{N} f_i^\dagger f_i\right] = (-1)^M \tag{1.2.15}$$

using the total number $M = \sum_{i=1}^{N} f_i^\dagger f_i$ of fermions. The minus sign on the right-hand side of (1.2.14) arises because in $K(N)$, $f_N^\dagger f_N$ is not contained. Defining

$$f_{N+1} = -f_1 \qquad (M : \text{even}) \tag{1.2.16a}$$

$$f_{N+1} = f_1 \qquad (M : \text{odd}) \tag{1.2.16b}$$

then (1.2.11) is valid as it stands.

As was mentioned in the previous section, (1.2.11) expresses competition between the formation of a spin singlet state and magnetic ordering, that is, competition between the kinetic energy of the fermions, i.e. itineracy, and the formation of density waves owing to the particle interaction. One remarkable feature of (1.2.11) is that in the case $J_z = 0$ (XY model); that is, in the quantum limit, the model becomes a free fermion theory that can be solved exactly. This limit is the opposite to the Ising limit that has been mentioned in the previous section. Introducing the Fourier transformations

$$f_n = \frac{1}{\sqrt{N}} \sum_k f_k \, \mathrm{e}^{\mathrm{i}kn} , \tag{1.2.17a}$$

$$f_n^\dagger = \frac{1}{\sqrt{N}} \sum_k f_k^\dagger \, \mathrm{e}^{-\mathrm{i}kn} , \tag{1.2.17b}$$

we obtain from (1.2.11)

$$H_{XY} = \sum_k \varepsilon(k) \cdot f_k^\dagger f_k \qquad (\varepsilon(k) = -J_\perp \cos k) . \tag{1.2.18}$$

The energy dispersion is shown in Fig. 1.3.

Owing to (1.2.1c), the relationship between the fermion number M and S_{tot}^z is $S_{\text{tot}}^z = M - N/2$. The Hilbertspace $S_{\text{tot}}^z = 0$ corresponds to the half-filled case $M = N/2$. In this case, the ground state is given by the state where the fermions occupy all states up to the Fermi energy $E_{\text{F}} = 0$. Excited states can be expressed as particle–hole creation. When N is even, the ground state is a singlet $S_{\text{tot}} = 0$, and the excited states start with $S_{\text{tot}} = 0$ or $S_{\text{tot}} = 1$.

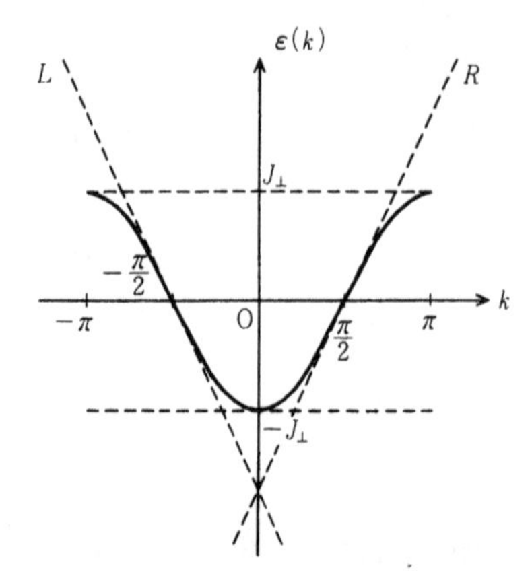

Fig. 1.3. The dispersion $\varepsilon(k) = J_\perp \cos k$. The dotted lines represent the linear approximation of the dispersion relation

The fact that this is expressed as particle–hole creation shows that fermions have spin $S = 1/2$.

Roughly speaking, one spin flip is realized by introducing two kinks, and therefore two fermions are necessary. In Fig. 1.4, the excitation energy of the particle–hole pair is shown as a function of $q = k_p - k_h$ (with k_p and k_h being the excitation energy of the particle and the hole, respectively).

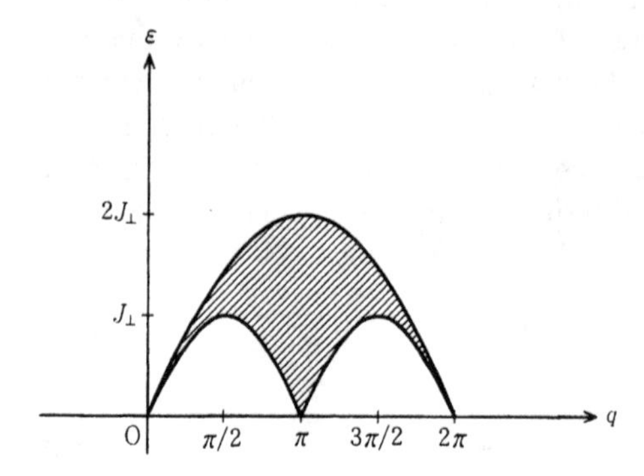

Fig. 1.4. The excitation of a particle–hole pair with wave number q

In such a manner, in the XY model, it is not the spin itself, but rather the kink – in the present case expressed as the Jordan–Wigner fermion – that is the elementary excitation, and the spin excitations can be expressed in terms of such combinations. In interacting particle systems – called many-body systems – in many cases new particles different from those that describe the original model are more fundamental. The quantum kink described here is such an example.

1.3 The Bethe Ansatz and the Exact Solution

In Sect. 1.1 and 1.2, we examined the classical limit ($J_\perp = 0$) and the quantum limit ($J_z = 0$) of the Hamiltonian (1.1.1). In fact, there exists an exact solution of the Hamiltonian (1.1.1) [(1.1.9)]. However, even when the wave function of the ground state is determined, this does not mean that all physical quantities can be calculated. Only by combining various theoretical approaches will it be possible to obtain a complete picture.

First, we start with the simple case of ferromagnetic interaction ($J_z < 0$). We indicate n_i as the site with spin $\uparrow$ at the ith side from the left. The wave function of the state where r spins are $\uparrow$ can be written as

$$\Phi_{n_1 n_2 \dots n_r} \,. \tag{1.3.1}$$

The total number of spins is given by N, therefore $N - r$ spins are $\downarrow$. The state Φ_0 where all spins are $\downarrow$ is an eigenfunction of (1.1.1):

$$H\Phi_0 = \frac{J_z}{4} N \Phi_0 \equiv E_0 \Phi_0 \,. \tag{1.3.2}$$

Here, we assume that the N spins are aligned as a one-dimensional ring (fulfilling periodic boundary conditions). Next, we consider the space of wave functions where just one spin is $\uparrow$, and with c_n being the coefficients of the linear superposition, we write

$$\Psi = \sum_{n=1}^{N} c_n \Phi_n \qquad (c_{n+N} = c_n) \,. \tag{1.3.3}$$

Acting with (1.1.1) on this state, we obtain

$$\begin{aligned} H\Psi &= \frac{J_z}{4}(N-4)\Psi + \frac{J_\perp}{2} \sum_{n=1}^{N} c_n (\Phi_{n-1} + \Phi_{n+1}) \\ &= \frac{J_z}{4}(N-4)\Psi + \frac{J_\perp}{2} \sum_{n=1}^{N} (c_{n+1} + c_{n-1}) \Phi_n \,. \end{aligned} \tag{1.3.4}$$

Writing $E\Psi$ for the right-hand side of (1.3.4) and comparing the coefficients of the Φ_n, we obtain

$$E c_n = (E_0 - J_z) c_n + \frac{J_\perp}{2}(c_{n+1} + c_{n-1}) \,. \tag{1.3.5}$$

This equation can be solved with plane waves

$$c_n = \frac{1}{\sqrt{N}} \mathrm{e}^{ikn} \,, \tag{1.3.6}$$

leading to the eigenvalues

$$E_k = E_0 - J_z + J_\perp \cos k\,. \tag{1.3.7}$$

Obviously, these are spin waves with wave number k. Owing to the boundary conditions $c_n = c_{n+N}$, the wave number is quantized as $k = (2\pi/N)m$ with m being an integer. The excitation energy $\varepsilon_k = E_k - E_0$ measured from the ground state energy is given by

$$\varepsilon_k = |J_z| + J_\perp \cos k\,, \tag{1.3.8}$$

and for $|J_z| \geq |J_\perp|$, ε_k is non-negative. However, for $|J_z| < |J_\perp|$, ε_k becomes negative, leading to a contradiction with the assumption made at the beginning that Φ_0 is the ground state. In this case, as described in Sect. 1.2, a spin liquid arises. In what follows, we assume that $|J_z| \geq |J_\perp|$.

Next, we consider the case where two spin waves are present ($r = 2$).

$$\Psi = \sum_{n_1<n_2} c_{n_1 n_2} \Phi_{n_1 n_2}\,. \tag{1.3.9}$$

Acting with H, it becomes clear that the case $n_2 = n_1 + 1$ differs from the other cases. That is,

$$\begin{aligned} H\Phi_{n_1 n_2} &= \left(E_0 - \frac{J_z}{2}\cdot 4\right)\Phi_{n_1 n_2} \\ &\quad + \frac{J_\perp}{2}\left(\Phi_{n_1-1,n_2} + \Phi_{n_1+1,n_2} + \Phi_{n_1,n_2-1} + \Phi_{n_1,n_2+1}\right) \\ &\quad (n_2 > n_1 + 1)\,. \end{aligned} \tag{1.3.10a}$$

$$H\Phi_{n_1,n_1+1} = \left(E_0 - \frac{J_z}{2}\cdot 2\right)\Phi_{n_1,n_1+1} + \frac{J_\perp}{2}\left(\Phi_{n_1-1,n_1+1} + \Phi_{n_1,n_1+2}\right). \tag{1.3.10b}$$

Transforming the problem again to an eigenvalue problem of the coefficients c_{n_1,n_2}, with $\varepsilon = E - E_0$, we obtain

$$\begin{aligned} \varepsilon c_{n_1 n_2} &= -2J_z c_{n_1 n_2} + \frac{J_\perp}{2}\left(c_{n_1+1,n_2} + c_{n_1-1,n_2} + c_{n_1,n_2+1} + c_{n_1,n_2-1}\right) \\ &\quad (n_2 > n_1 + 1) \end{aligned} \tag{1.3.11a}$$

$$\varepsilon c_{n_1,n_1+1} = -J_z c_{n_1,n_1+1} + \frac{J_\perp}{2}\left(c_{n_1-1,n_1+1} + c_{n_1,n_1+2}\right). \tag{1.3.11b}$$

First, we study the case when $n_2 > n_1 + 1$. We set

$$c_{n_1 n_2} = c_1\, \mathrm{e}^{\mathrm{i}(k_1 n_1 + k_2 n_2)} + c_2\, \mathrm{e}^{\mathrm{i}(k_2 n_1 + k_1 n_2)}\,. \tag{1.3.12}$$

This is the expression of two propagating spin waves with wave number k_1 and k_2. Here, because the restriction $n_2 > n_1$ has been imposed, in contrast to the first term where the spin wave with wave number k_1 runs behind k_2, the second term can be seen as the part where k_1 'surpassed' k_2. In general, when this 'surpassing' arises, energy and momentum transfer might occur. In a single dimension, the equations

$$k_1 + k_2 = k_3 + k_4 , \tag{1.3.13a}$$

$$\varepsilon_{k_1} + \varepsilon_{k_2} = \varepsilon_{k_3} + \varepsilon_{k_4} \tag{1.3.13b}$$

are only satisfied for k_3 and k_4 with $(k_3, k_4) = (k_1, k_2)$ or (k_2, k_1). Therefore, equation (1.3.12) holds. So, how does the spin wave scattering during the 'surpassing' process show up? The answer is a 'phase shift'. Scattering arises when the spins are neighbouring, as described by (1.3.11b). We consider now the case when $n_1 \leq n_2 \leq n_1 + 1$. To do so, we perform a little trick. $c_{n_1 n_2}$ in equation (1.3.12) as it stands is defined for $n_1 < n_2$. Generalizing also for the case $n_1 = n_2$, $c_{n_1 n_1}$ can be defined in the following manner. Assuming that (1.3.11b) is consistent with (1.3.11a) in this case, (1.3.12) as it stands is the solution to the eigenvalue problem. Comparing (1.3.11a) with (1.3.11b), the condition that determines $c_{n_1 n_1}$ is given by

$$-J_z c_{n_1, n_1+1} + \frac{J_\perp}{2}(c_{n_1,n_1} + c_{n_1+1,n_1+1}) = 0 . \tag{1.3.14}$$

Under this condition, (1.3.11a) is valid for all n_1 and n_2; and (1.3.12) corresponds to the eigenstate with eigenvalue

$$\varepsilon = \varepsilon_{k_1} + \varepsilon_{k_2} . \tag{1.3.15}$$

Then, for c_1 and c_2 (1.3.14) leads to the condition:

$$-2J_z\left(c_1\,\mathrm{e}^{\mathrm{i}k_2} + c_2\,\mathrm{e}^{\mathrm{i}k_1}\right) + J_\perp (c_1 + c_2)\left(1 + \mathrm{e}^{\mathrm{i}(k_1+k_2)}\right) = 0 . \tag{1.3.16}$$

This equals

$$\frac{c_1}{c_2} = -\frac{J_z\,\mathrm{e}^{\mathrm{i}\frac{k_1-k_2}{2}} - J_\perp \cos\dfrac{k_1+k_2}{2}}{J_z\,\mathrm{e}^{-\mathrm{i}\frac{k_1-k_2}{2}} - J_\perp \cos\dfrac{k_1+k_2}{2}} . \tag{1.3.17}$$

Because the numerator is the complex conjugate of the denominator, $|c_1/c_2| = 1$. Introducing the 'phase shift' ϕ, we can write

$$c_1 = \mathrm{e}^{\mathrm{i}\phi/2} , \qquad c_2 = \mathrm{e}^{-\mathrm{i}\phi/2} . \tag{1.3.18}$$

Then, (1.3.16) can be expressed as

$$\cot\frac{\phi}{2} = \frac{J_z \sin\dfrac{k_1-k_2}{2}}{J_\perp \cos\dfrac{k_1+k_2}{2} - J_z \cos\dfrac{k_1-k_2}{2}} . \tag{1.3.19}$$

This equation determines ϕ. Choosing the region $|\phi| < \pi$, and considering the case of the Heisenberg model $J_\perp = J_z$, (1.3.19) becomes

$$\cot\frac{\phi}{2} = \frac{\sin\frac{k_1}{2}\cos\frac{k_2}{2} - \cos\frac{k_1}{2}\sin\frac{k_2}{2}}{-2\sin\frac{k_1}{2}\sin\frac{k_2}{2}} = \frac{1}{2}\left(\cot\frac{k_1}{2} - \cot\frac{k_2}{2}\right). \tag{1.3.20}$$

Imposing periodic boundary conditions

$$c_{n_1 n_2} = c_{n_2, n_1+N}, \tag{1.3.21}$$

we obtain from (1.3.12)

$$\mathrm{e}^{\mathrm{i}\phi/2} = \mathrm{e}^{\mathrm{i}k_1 N - \mathrm{i}\phi/2}, \qquad \mathrm{e}^{-\mathrm{i}\phi/2} = \mathrm{e}^{\mathrm{i}k_2 N + \mathrm{i}\phi/2}. \tag{1.3.22}$$

We conclude that with $m_1, m_2 = 0, 1, 2, \ldots, N-1$, we obtain

$$k_1 N = \phi + 2\pi m_1 \tag{1.3.23a}$$

$$k_2 N = -\phi + 2\pi m_2. \tag{1.3.23b}$$

Comparing this with the wave number $k = 2\pi m/N$ of a single spin wave introduced below (1.3.7), we conclude from (1.3.23) that the phase shift is just given by $\pm\phi/N$. In this manner, the interaction of the spin waves shows up.

Having fixed m_1 and m_2, by (1.3.23), k_1 and k_2 can be expressed only with ϕ. We insert this expression into (1.3.20). In this manner, an equation for ϕ is obtained. Fig. 1.5 shows ϕ as a function of $\cot(\phi/2)$. Because now, m_1 and m_2 can be interchanged, we set $m_1 \leq m_2$. Fixing k_2, when k_1 changes from 0 to k_2, the right-hand side of (1.3.20) decreases from ∞ to zero, and therefore ϕ changes from zero to π, and m_1 from zero to $m_2 - 1$. For $m_1 = m_2 - 1$, since $\phi = \pi$, the relationship $k_1 = k_2 = k$ holds. However, because of

$$c_{n_1 n_2} = \mathrm{e}^{\mathrm{i}\phi/2 + \mathrm{i}k(n_1+n_2)} + \mathrm{e}^{-\mathrm{i}\phi/2 + \mathrm{i}k(n_1+n_2)} = 2\,\mathrm{e}^{\mathrm{i}k(n_1+n_2)}\cos\frac{\phi}{2} = 0, \tag{1.3.24}$$

in this case a solution does not exist.

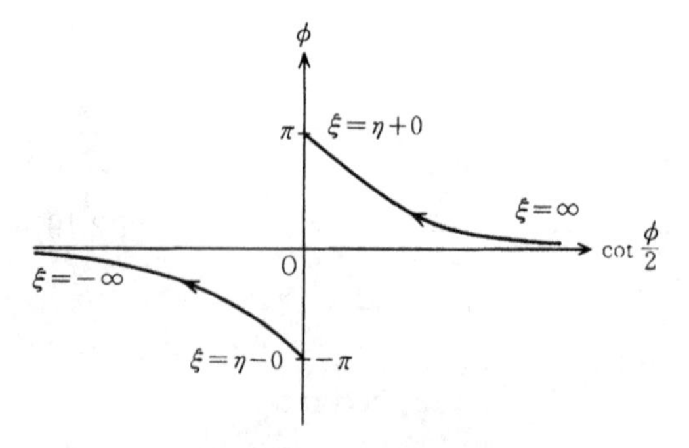

Fig. 1.5. ϕ as a function of $\cot\phi/2$. ϕ is restricted to the interval $[-\pi, \pi]$. The relation to ξ is explained in the text below (1.3.48)

On the other hand, for the $m_2 - 1$ values $m_1 = 0, 1, \ldots, m_2 - 2$, c_{n_1,n_2} is finite and can be used. The solution for the excitation energy as a function of the total wave number $K = k_1 + k_2$ with real k_1 and k_2 leads to a dispersion relation with continuous spectrum. That is, with $q = k_1 - k_2$, the excitation energy of two spin waves is given by

$$\varepsilon(K, q) = \varepsilon_{\frac{K+q}{2}} + \varepsilon_{\frac{K-q}{2}} = 2|J_z| \left(1 - \cos\frac{K}{2}\cos\frac{q}{2}\right). \tag{1.3.25}$$

For every K, the spectrum is continuous with a half width of $4\,|J_z|\cos K/2$.

Now, is this the complete solution? To see this, we count the number of solutions. The number of the above scattering solutions is

$$N_\mathrm{s} = \sum_{m_2=2}^{N-1}(m_2 - 1) = \frac{(N-1)(N-2)}{2}. \tag{1.3.26}$$

Because the number of independent solutions is

$$N_\mathrm{t} = \sum_{m_2=1}^{N-1} m_2 = \frac{N(N-1)}{2}, \tag{1.3.27}$$

there are still $N_\mathrm{t} - N_\mathrm{s} = N - 1$ solutions missing. These remaining solutions correspond to bound states of two spin waves. In the limit $|J_z| \gg |J_\perp|$, the existence of bound state solutions should be evident. The energy cost expressed in the J_z term is reduced by half when two $\uparrow$ spins are next to each other.

However, for the case $|J_z| = |J_\perp|$ of the Heisenberg model, this is not evident. Because for a bound state the wave function should exponentially decay with the distance between the two $\uparrow$ spins, we expect that $k_1 - k_2$ is purely imaginary. On the other hand, because the centre of mass movement should be a plane wave, $K = k_1 + k_2$ should still be real. With real u and v, we write

$$\begin{aligned} k_1 &= u + \mathrm{i}v\,, \\ k_2 &= u - \mathrm{i}v\,. \end{aligned} \tag{1.3.28}$$

Then, (1.3.23) becomes

$$\begin{aligned} u &= \frac{\pi}{N}(m_1 + m_2)\,, \\ \phi &= \pi(m_2 - m_1) + \mathrm{i}Nv\,. \end{aligned} \tag{1.3.29}$$

Inserting (1.3.29) into (1.3.17) (and setting $J_z = J_\perp$), we obtain

$$\mathrm{e}^{\mathrm{i}\pi(m_2 - m_1)}\,\mathrm{e}^{-Nv} = -\frac{\cos u - \mathrm{e}^{-v}}{\cos u - \mathrm{e}^{v}}. \tag{1.3.30}$$

Performing the limit $N \to \infty$ with v resting finite, we obtain

$$\mathrm{e}^{-v} = \cos u\,.$$

When $u = K/2$ is fixed, this equation determines v. As becomes clear when (1.3.28) is inserted into (1.3.12), v is the inverse of the localization length for the relative coordinate $n_1 - n_2$. The energy of the bound state can be calculated as

$$\begin{aligned}\varepsilon_{\mathrm{B}}(K) &= 2|J_z| - |J_z|\{\cos(u+\mathrm{i}v) + \cos(u-\mathrm{i}v)\} \\ &= 2|J_z|(1-\cos u \cosh v) \\ &= 2|J_z|\left(1-\frac{\cos^2 u + 1}{2}\right) = |J_z|\left(1-\cos^2\frac{K}{2}\right) \\ &= \frac{1}{2}|J_z|(1-\cos K)\,. \end{aligned} \quad (1.3.31)$$

Above, the ferromagnetic spin wave has been considered. Next, we discuss the ground state and excitations in the anti-ferromagnetic Heisenberg model ($J_z = J_\perp = J > 0$). In this case, $r = N/2$ spins are ↑ in the ground state. That is, the ground state solution does exist in the subspace $S^z_{\mathrm{tot}} = 0$.

$$\Psi = \sum_{n_1<n_2<\cdots<n_r} c_{n_1 n_2 \dots n_r} \Phi_{n_1 n_2 \dots n_r} \qquad \left(r = \frac{N}{2}\right). \quad (1.3.32)$$

Now, we try for $c_{n_1 n_2 \dots n_r}$ the following Ansatz, which is called Bethe Ansatz:

$$c_{n_1 n_2 \dots n_r} = \sum_P \exp\left[\mathrm{i}\left(\sum_{j=1}^{r} k_{P_j} n_j + \frac{1}{2}\sum_{j<1} \phi_{P_j,P_l}\right)\right]. \quad (1.3.33)$$

Here, P is the permutation operator of the numbers $1\dots r$, and $\phi_{j,l}$ is the phase shift, satisfying $\phi_{j,l} = \phi_{l,j}$. Furthermore, we impose periodic boundary conditions for $c_{n_1 n_2 \dots n_r}$:

$$c_{n_1 n_2 \dots n_r} = c_{n_2 n_3 \dots n_r, n_1+N}\,. \quad (1.3.34)$$

(1.3.33) is the generalization of (1.3.12). It has been assumed that the scattering of ↑ only interchanges the wave numbers $k_1 k_2 \dots k_r$, leading to a new wave function, and that many-particle scattering can be expressed as a product of two-body scattering. Indeed, when acting with H on Ψ (1.3.32), applying the same trick as when (1.3.32) was deduced, we obtain

$$2c_{n_1,\dots,n_k,n_{k+1},\dots,n_r} = c_{n_1,\dots,n_k,n_k,\dots,n_r} + c_{n_1,\dots,n_{k+1},n_{k+1},\dots,n_r} \quad (1.3.35)$$

and can prove that

$$H\Psi = E\Psi\,. \quad (1.3.36)$$

In this case, the energy eigenvalue is given by

$$E = -\frac{NJ}{4} + J\sum_{j=1}^{r}\cos k_j = \frac{NJ}{4} - J\sum_{j=1}^{r}(1-\cos k_j)\,. \quad (1.3.37)$$

Performing a calculation similar to the derivation of (1.3.20), we obtain from (1.3.35)

$$\cot\frac{\phi_{j,l}}{2} = \frac{1}{2}\left(\cot\frac{k_j}{2} - \cot\frac{k_l}{2}\right). \tag{1.3.38}$$

On the other hand, the right-hand side of the periodic boundary condition (1.3.34) becomes

$$\begin{aligned} c_{n_2 n_3 \ldots n_r, n_1+N} &= \sum_P \exp\left[\mathrm{i}\sum_{j=1}^{r} k_{P_j} n_{j+1} + \mathrm{i}k_{P_r}(n_1+N) + \frac{\mathrm{i}}{2}\sum_{j<1}\phi_{P_j,P_l}\right] \\ &= \sum_{P'} \exp\left[\mathrm{i}\sum_{j=2}^{r-1} k_{P'_{j-1}} n_j + \mathrm{i}k_{P'_r}(n_1+N) + \frac{\mathrm{i}}{2}\sum_{j<l}\phi_{P'_j,P'_l}\right]. \end{aligned} \tag{1.3.39}$$

Here, we substituted the sum over P by P'. Because the left-hand side of equation (1.3.34) is (1.3.33), there must be a correspondence between this summation in P and the summation in P' appearing at the far right-hand side of (1.3.39):

$$\begin{aligned} P'_{j-1} &= P_j\,, \\ P'_r &= P_1 \qquad (r \geqq j \geqq 2)\,, \end{aligned} \tag{1.3.40}$$

for every single term. We obtain the condition

$$Nk_{P'_r} + \frac{1}{2}\sum_{j<l}\phi_{P'_j,P'_j} - \frac{1}{2}\sum_{j<l}\phi_{P_j,P_l} = 2\pi m_{P'_r}\,. \tag{1.3.41}$$

We look at the left-hand side of this equation a bit more in detail.

$$\begin{aligned} &Nk_{P_1} + \frac{1}{2}\sum_{j<l\leqq r-1}\phi_{P_{j+1},P_{l+1}} + \frac{1}{2}\sum_{j=1}^{r-1}\phi_{P_{j+1},P_1} - \frac{1}{2}\sum_{2\leqq j<1}\phi_{P_j,P_l} - \frac{1}{2}\sum_{l=2}^{r}\phi_{P_1,P_l} \\ &= Nk_{P_1} - \sum_{l=2}^{r}\phi_{P_1,P_l} = 2\pi m_{P_1}\,. \end{aligned} \tag{1.3.42}$$

Because P_1 is variable, (1.3.42) can be expressed as

$$Nk_j = \sum_{l\neq j}\phi_{j,l} + 2\pi m_j\,. \tag{1.3.43}$$

The meaning of this equation is obvious. When one particle goes around the ring, it scatters once with every particle, and the resulting total phase shift is given by the first term at the right-hand side.

Next, the state is determined by a combination $\{m_i\}$ of integer numbers, and the ground state is given by

$$\{m_j\} = \{1, 3, 5, 7, \ldots, N-1\}. \tag{1.3.44}$$

This is owing to the fact that two neighbouring m correspond to an anti-bound state, as follows from the ferromagnetic spin wave theory developed above. That is, in the case of anti-ferromagnetic interaction, the energy becomes larger when an anti-binding state of two neighbouring $\uparrow$ spins arises.

Next, we consider the thermodynamic limit $N \to \infty$ and introduce continuous variables x and y in (1.3.43):

$$x = \frac{2j-1}{N}, \qquad y = \frac{2l-1}{N}. \tag{1.3.45}$$

Recalling $m_j = 2j - 1$ and dividing by N on both sides in (1.3.43), we obtain

$$k(x) = 2\pi x + \frac{1}{2}\int_0^1 \mathrm{d}y\, \phi(k(x), k(y)). \tag{1.3.46}$$

Here, owing to (1.3.38), $\phi(k(x), k(y))$ is given by

$$\cot\frac{\phi(k(x), k(y))}{2} = \frac{1}{2}\left(\cot\frac{k(x)}{2} - \cot\frac{k(y)}{2}\right), \tag{1.3.47}$$

and from (1.3.37) we conclude that the energy is given by

$$\begin{aligned} E &= \frac{NJ}{4} - J\sum_{j=1}^{N/2}[1 - \cos k_j] \\ &= \frac{NJ}{4} - \frac{NJ}{2}\int_0^1 \mathrm{d}x\,[1 - \cos k(x)]. \end{aligned} \tag{1.3.48}$$

Setting $\xi = \cot k(x)/2$, $k(0) = 0$ corresponds to $\xi = +\infty$, and $k(1) = 2\pi$ corresponds to $\xi = -\infty$. As shown in Fig. 1.5 and given by (1.3.47), the transformation $\phi(k(x), k(y)) = \phi(\xi, \eta)$ with $\eta \equiv \cot k(y)/2$ leads to the identification $\phi(\xi, \eta) = +0$ at $\xi = +\infty$, $\phi(\xi, \eta) = \pi$ at $\xi = \eta + 0$, $\phi(\xi, \eta) = -\pi$ at $\xi = \eta - 0$, and $\phi(\xi, \eta) = -0$ at $\xi = -\infty$. Writing (1.3.46) as

$$k(x) = 2\pi x + \int_0^1 \cot^{-1}\frac{\xi(x) - \eta(y)}{2}\,\mathrm{d}y \tag{1.3.49}$$

we conclude that $\cot^{-1}(\xi - \eta)/2$ jumps at $\xi = \eta$ from $+\pi$ to $-\pi$. Differentiating both sides from (1.3.49) with respect to x, we obtain

$$\begin{aligned} \frac{\mathrm{d}k(x)}{\mathrm{d}x} &= 2\pi + \int_{+\infty}^{-\infty}\left[\frac{-1}{1 + (\xi - \eta)^2/4} + 2\pi\delta(\xi - \eta)\right]\frac{1}{2}\frac{\mathrm{d}\xi}{dx}\frac{\mathrm{d}y}{\mathrm{d}\eta}\,\mathrm{d}\eta \\ &= \pi + \int_{-\infty}^{+\infty}\frac{2}{4 + (\xi - \eta)^2}\frac{\mathrm{d}y}{\mathrm{d}\eta}\,\mathrm{d}\eta \cdot \frac{\mathrm{d}\xi}{\mathrm{d}x}. \end{aligned} \tag{1.3.50}$$

Introducing the function

$$\frac{\mathrm{d}\xi}{\mathrm{d}x} = -\frac{1}{g(\xi)} \tag{1.3.51}$$

owing to

$$\frac{\mathrm{d}\xi(x)}{\mathrm{d}x} = \frac{\mathrm{d}}{\mathrm{d}x}\left[\cot\frac{k(x)}{2}\right] = -\frac{1}{2}\frac{1}{\sin^2\dfrac{k(x)}{2}}\frac{\mathrm{d}k(x)}{\mathrm{d}x}$$
$$= -\frac{1}{2}\left(1+\cot^2\frac{k(x)}{2}\right)\frac{\mathrm{d}k(x)}{\mathrm{d}x} = -\frac{1}{2}\left(1+\xi(x)^2\right)\frac{\mathrm{d}k(x)}{\mathrm{d}x},$$

we obtain

$$\frac{\mathrm{d}k(x)}{\mathrm{d}x} = -\frac{2}{1+\xi^2}\frac{\mathrm{d}\xi}{\mathrm{d}x} = \frac{2}{1+\xi^2}\frac{1}{g(\xi)}.$$

Using this, (1.3.50) can be re-expressed as

$$\frac{2}{1+\xi^2} = \pi g(\xi) + 2\int_{-\infty}^{\infty}\frac{g(\eta)}{4+(\xi-\eta)^2}\,\mathrm{d}\eta. \tag{1.3.52}$$

Introducing the Fourier transformation

$$G(u) = \int_{-\infty}^{\infty} g(\xi)\,\mathrm{e}^{\mathrm{i}\xi u}\,\mathrm{d}\xi, \tag{1.3.53}$$

we obtain from (1.3.52) the result

$$G(u) = \frac{1}{\cosh u}. \tag{1.3.54}$$

We now determine the energy. From (1.3.48), we deduce that

$$E - \frac{NJ}{4} = -\frac{NJ}{2}\int_0^1 \mathrm{d}x\,(1-\cos k(x)) = -JN\int_0^1 \mathrm{d}x\sin^2\frac{k(x)}{2}$$
$$= -JN\int_{+\infty}^{-\infty}\frac{1}{1+\cot^2\dfrac{k(x)}{2}}\frac{\mathrm{d}x}{\mathrm{d}\xi}\,\mathrm{d}\xi = -JN\int_{-\infty}^{\infty}\frac{g(\xi)}{1+\xi^2}\,\mathrm{d}\xi$$
$$= -JN\int_{-\infty}^{\infty}\frac{\mathrm{d}\xi}{2\pi}\int_{-\infty}^{\infty}\mathrm{d}u\,\frac{\mathrm{e}^{-\mathrm{i}\xi u}}{1+\xi^2}G(u)$$
$$= -\frac{JN}{2}\int_{-\infty}^{\infty}\mathrm{d}u\,\frac{\mathrm{e}^{-|u|}}{\cosh u} = -JN\ln 2. \tag{1.3.55}$$

Above, we developed the Bethe Ansatz for the ground state. Next, we discuss low energy excitations. Now, we alter the combination of $\{m_i\}$ slightly:

$$\{m_j\} = \{0, 2, 4, \ldots, (2n-2), (2n+1), \ldots, (N-1)\}, \tag{1.3.56}$$

by shifting the first n integers m_j one step to the left. Then, the total momentum is given by

$$K = \sum_j k_j = \frac{2\pi}{N} \sum_j m_j . \tag{1.3.57}$$

Calling q the difference momentum measured from the ground state, we obtain

$$q = \frac{2\pi}{N} \sum_{j=1}^{n} [(2j-2) - (2j-1)] = -\frac{2\pi n}{N} . \tag{1.3.58}$$

Performing the continuum limit and setting $x = (2j-1)/N$, we introduce $\lambda(x)$ as

$$\lambda(x) = x - \frac{1}{N}\theta\left(\frac{|q|}{\pi} - x\right) . \tag{1.3.59}$$

The second term on the right-hand side corresponds to the shift of n integers m_j one step to the left. Then, instead of (1.3.49), we obtain

$$k(x) = 2\pi\lambda(x) + \int_0^1 \cot^{-1}\frac{\xi - n}{2}\,dy , \tag{1.3.60}$$

and instead of (1.3.52) we obtain

$$\frac{2}{1+\xi^2} - \frac{2\pi}{N}\delta(\xi - \xi_0) = \pi g(\xi) + 2\int_{-\infty}^{\infty} \frac{g(\eta)}{4 + (\xi - \eta)^2}\,\mathrm{d}\eta . \tag{1.3.61}$$

Here, ξ_0 is given by $\xi_0 = \cot \frac{1}{2} k(|q|/\pi)$.

The $o(\frac{1}{N})$ order correction on the left-hand side leads to a correction of the same order for $g(\xi)$ and $G(u)$, and an excitation energy of order $o(1)$ emerges. This will be explained in what follows. From (1.3.61), the Fourier transformation $G(u)$ is determined to be

$$G(u) = \frac{1}{\cosh u} - \frac{2}{N} \cdot \frac{\mathrm{e}^{iu\xi_0}}{1 + \mathrm{e}^{-2|u|}} . \tag{1.3.62}$$

From this equation, we obtain

$$\begin{aligned} \Delta E &= -NJ \int_{-\infty}^{\infty} \frac{\Delta g(\xi)}{1+\xi^2}\,\mathrm{d}\xi = +\frac{J}{2}\int_{-\infty}^{\infty} \mathrm{d}u\, \frac{\mathrm{e}^{iu\xi_0}}{\cosh u} \\ &= \frac{\pi J/2}{\cosh(\pi\xi_0/2)} . \end{aligned} \tag{1.3.63}$$

Now, only the relationship between ξ_0 and q has to be determined. From (1.3.51) we obtain

$$x(\xi_0) - x(-\infty) = -\int_{-\infty}^{\xi_0} g(\xi)\,\mathrm{d}\xi \tag{1.3.64}$$

and it is sufficient to use $g(\xi)$ as given in (1.3.54) without higher order corrections. Performing the Fourier transformation in (1.3.54), we obtain

$$g(\xi) = \frac{1}{2\cosh\dfrac{\pi\xi}{2}}\,. \qquad (1.3.65)$$

Owing to $x(\xi = -\infty) = 1$, it follows that

$$\begin{aligned} x(\xi_0) &= 1 - \int_{-\infty}^{\xi_0} \frac{d\xi}{2\cosh\frac{\pi\xi}{2}} = 1 - \int_{-\infty}^{y_0} \frac{dy}{\pi(1+y^2)} \\ &= \frac{1}{\pi}\cot^{-1}\left(\sinh\frac{\pi\xi_0}{2}\right)\,. \end{aligned} \qquad (1.3.66)$$

Here, the substitution $y = \sinh \pi\xi/2$ and the boundary condition $\cot^{-1}(-\infty) = -0$ for $\xi_0 \to -\infty$ have been used. By taking the cot on both sides of the equation

$$x(\xi_0) = \frac{|q|}{\pi} = \frac{1}{\pi}\cot^{-1}\left(\sinh\frac{\pi\xi_0}{2}\right)\,, \qquad (1.3.67)$$

we obtain

$$\cot|q| = \sinh\frac{\pi\xi_0}{2}\,, \qquad (1.3.68)$$

and therefore

$$\cosh\frac{\pi\xi_0}{2} = \sqrt{1+\left(\sinh\frac{\pi\xi_0}{2}\right)^2} = \sqrt{1+\cot^2|q|} = \frac{1}{\sin|q|}\,.$$

Finally, we obtain the result

$$\Delta E = \frac{\pi J}{2}|\sin q|\,. \qquad (1.3.69)$$

This is the scattering relation of the so-called des Cloiseaux–Pearson excitation mode. This mode behaves like the spin wave excitations which would appear when anti-ferromagnetic long-range ordering is present. That is, at $q = 0$ and $q = \pi$, the dispersion is linear, and at $q = \pm\pi/2$, the Brioullin zone is folded. However, notice that neither long-range ordering, nor the symmetry breaking occurs. Rather than to the dispersion relation of spin waves, (1.3.69) corresponds to the lower bound of the continuum spectrum shown in Fig. 1.2 and Fig. 1.4. In Fig. 1.2, with increasing $J_\perp$, the oblique region becomes larger, and for $|J_\perp| = J_z$, the lowest boundary of the energy spectrum becomes zero for $q = 0$ and $q = \pm\pi$. In the limit $|J_\perp|/J_z \to \infty$, the excitation spectrum becomes similar to Fig. 1.4.

2. Quantum Field Theory in 1+1 Dimensions

There exist many powerful theoretical approaches to the description of one-dimensional, many-particle quantum systems. In this chapter, we choose the one-dimensional quantum spin system as an example and analyse it in the framework of quantum field theory.

2.1 Bosonization

In this chapter, the fermion model derived in Sect. 1.2 will be discussed further. Interacting one-dimensional, many-particle quantum liquids are described by so-called Tomonaga–Luttinger liquids, and the XXZ model is an example.

First, we introduce the continuum limit of the Hamiltonian (1.2.11). Naively, we might assume that for the case when J_z is small compared with $J_\perp$, the kinetic energy of the fermions is dominant, and the interaction is weak, so that perturbation theory can be applied. To be precise, this is not totally correct; anyhow, in the band structure described by $J_\perp$ in (1.2.18), J_z creates particle–hole pairs, where the low-energy particle–hole pairs in the vicinity of the Fermi surface should be dominant.

Now, we replace the dispersion relation in the vicinity of $k = \pm k_\mathrm{F} = \pm\pi/2$ by two linearized fermion dispersion relations. Then, the fermion annihilation operator f_n reads

$$f_n = R(x_n)\,\mathrm{e}^{\mathrm{i}k_\mathrm{F}x_n} + L(x_n)\,\mathrm{e}^{-\mathrm{i}k_\mathrm{F}x_n}\,. \tag{2.1.1}$$

Here, x_n is given by $x_n = n$, and because $R(x)$ and $L(x)$ contain only long wavelength components that almost do not change in about one lattice spacing $a = 1$, in what follows we assume x to be a continuous coordinate. In doing so, the Hamiltonian (1.2.11) can be expressed in the following compact manner:

$$H_{XY} = -J_\perp \int \mathrm{d}x\,\bar{\psi}\,\mathrm{i}\gamma_1\partial_x\psi\,. \tag{2.1.2}$$

Here, we introduced some new notations. First, we defined the two-component spinor:

$$\psi(x) = \begin{pmatrix} R(x) \\ L(x) \end{pmatrix} . \tag{2.1.3}$$

On the other hand, $\bar{\psi}$ is defined as

$$\bar{\psi}(x) = \left(L^\dagger(x), R^\dagger(x)\right) . \tag{2.1.4}$$

Notice that the order of R and L in (2.1.3) is opposite to (2.1.4). Next, we introduced the Dirac matrices γ_μ in $1+1$ dimensions as

$$\gamma_0 = \sigma_x = \begin{bmatrix} 0 & 1 \\ 1 & 0 \end{bmatrix} \tag{2.1.5a}$$

$$\gamma_1 = -\mathrm{i}\sigma_2 = \begin{bmatrix} 0 & -1 \\ 1 & 0 \end{bmatrix} \tag{2.1.5b}$$

$$\gamma_5 = \gamma_0\gamma_1 = \sigma_3 = \begin{bmatrix} 1 & 0 \\ 0 & -1 \end{bmatrix} . \tag{2.1.5c}$$

Then, $\bar{\psi} = \psi^\dagger\gamma_0$ holds, and we obtain

$$\begin{aligned} \bar{\psi}\,\mathrm{i}\gamma_1\partial_x\psi &= \left(L^\dagger, R^\dagger\right)\begin{pmatrix} 0 & -\mathrm{i} \\ \mathrm{i} & 0 \end{pmatrix}\begin{pmatrix} \partial_x R \\ \partial_x L \end{pmatrix} \\ &= \mathrm{i}\left(R^\dagger\partial_x R - L^\dagger\partial_x L\right) . \end{aligned} \tag{2.1.6}$$

Performing a Fourier transformation, obviously, (2.1.2) becomes

$$H_{XY} = \sum_k J_\perp k\left(R^\dagger(k)R(k) - L^\dagger(k)L(k)\right) . \tag{2.1.7}$$

Owing to (2.1.1), the term proportional to J_z becomes

$$\begin{aligned} f_n^\dagger f_n &= \left[R^\dagger(x_n)R(x_n) + L^\dagger(x_n)L(x_n)\right] \\ &\quad + \mathrm{e}^{2\mathrm{i}k_\mathrm{F}x_n}\left[R^\dagger(x_n)L(x_n) + L^\dagger(x_n)R(x_n)\right] , \end{aligned} \tag{2.1.8}$$

and owing to $\mathrm{e}^{2\mathrm{i}k_\mathrm{F}x_n} = (-1)^n$, the first term on the right-hand side corresponds to the uniform component, and the second to the alternating (staggered) component.

Next, we introduce the normal product. Generally, let f_k and $f_k^\dagger$ be the fermion operators, then at $T = 0$, the states $|k| < k_\mathrm{F}$ are occupied, and the states $|k| > k_\mathrm{F}$ are not. Writing $|F\rangle$ for the state where all states are occupied up to the Fermi surface, then, for example the normal product $: f_{k_1}^\dagger f_{k_2} :$ of $f_{k_1}^\dagger f_{k_2}$ is defined as

$$: f_{k_1}^\dagger f_{k_2} : = \begin{cases} f_{k_1}^\dagger f_{k_2} & (|k_2| > k_\mathrm{F}) \\ -f_{k_2} f_{k_1}^\dagger & (|k_2| < k_\mathrm{F}) \end{cases} . \tag{2.1.9}$$

That is, operators that annihilate $|F\rangle$ are placed on the right of the other operators, taking care of the fermionic anti-commutation relations. Equation (2.1.9) can be expressed as

$$:f_{k_1}^\dagger f_{k_2}: \; = f_{k_1}^\dagger f_{k_2} - \langle F|f_{k_1}^\dagger f_{k_2}|F\rangle \,. \tag{2.1.10}$$

Applying this to $f_n^\dagger f_n$ in (2.1.8), owing to $\langle F|f_n^\dagger f_n|F\rangle = 1/2$, we obtain

$$\begin{aligned} f_n^\dagger f_n - \tfrac{1}{2} &= \; :f_n^\dagger f_n: \\ &= \left[:R^\dagger(x_n)R(x_n): + :L^\dagger(x_n)L(x_n): \right] \\ &\quad + \mathrm{e}^{\mathrm{i}\pi n}\left[:R^\dagger(x_n)L(x_n): + :L^\dagger(x_n)R(x_n): \right] . \end{aligned} \tag{2.1.11}$$

We now mention a subtle and important feature of the continuum limit. When the dispersion of $R, R^\dagger$ or $L, L^\dagger$ extends over the region $k = -\infty$ to $k = +\infty$, as shown in Fig. 1.3, then, for example,

$$\begin{aligned} \langle F|R^\dagger(x)R(x)|F\rangle &= \frac{1}{L}\sum_k \langle F|R^\dagger(k)R(k)|F\rangle \\ &= \frac{1}{L}\sum_{k<0} 1 \end{aligned} \tag{2.1.12}$$

diverges. When $R^\dagger(x)R(x)$ or $L^\dagger(x)L(x)$ are used directly, the problem of infinities arises. Therefore, we use the normal product introduced in (2.1.11)

$$\rho_\mathrm{R}(x) = \; :R^\dagger(x)R(x): \tag{2.1.13a}$$

$$\rho_\mathrm{L}(x) = \; :L^\dagger(x)L(x): \tag{2.1.13b}$$

describing the fluctuation of the right-moving and left-moving fermion densities, measured from the state $|F\rangle$. Now, bearing this fact in mind, we will omit the notation : : in the following, except for cases where it becomes necessary. The principal idea of bosonization is to describe the fermionic system in terms of the density fluctuation operator in (2.1.13) (being a bosonic operator because it is the product of two fermionic operators). Writing $\rho(x)$ for the density of the total system and $j(x)$ for the current, we obtain

$$\rho(x) = \rho_\mathrm{R}(x) + \rho_\mathrm{L}(x) \,, \tag{2.1.14a}$$

$$j(x) = \rho_\mathrm{R}(x) - \rho_\mathrm{L}(x) \,. \tag{2.1.14b}$$

Another possible interpretation of $\rho_\mathrm{R}(x)$ and $\rho_\mathrm{L}(x)$ is the shift of the Fermi point. Within the region of the order of k_F^{-1}, the spatial dependence of $\rho_\mathrm{R}(x), \rho_\mathrm{L}(x)$ can be ignored, and therefore the Fermi wave numbers $k_\mathrm{F}^\mathrm{R}(x)$ and $k_\mathrm{F}^\mathrm{L}(x)$ are shifted around the point x:

$$k_\mathrm{F}^\mathrm{R}(x) = k_\mathrm{F} + 2\pi\rho_\mathrm{R}(x) \,, \tag{2.1.15a}$$

$$k_\mathrm{F}^\mathrm{L}(x) = -k_\mathrm{F} - 2\pi\rho_\mathrm{L}(x) \,. \tag{2.1.15b}$$

That is, $\rho_{\rm R}(x)$ and $\rho_{\rm L}(x)$ express the space-dependent deformation of the Fermi sphere, which in a single dimension is exhausted by the collective modes because the 'Fermi surface' consists only of two points. This is the basic point for the one-dimensional bosonization, and this point will become important again later when bosonization of higher dimensional systems is discussed.

Now, we discuss the properties of the operators $\rho_{\rm R}(x), \rho_{\rm L}(x)$. The most basic property is the commutation relation, and, at first sight, the reader might think that $\rho_R(x)$ and $\rho_R(x')$ ($\rho_L(x)$ and $\rho_L(x')$) do commute. However, as has been mentioned above, when the dispersion of the whole region $-\infty$ to $+\infty$ is taken into account, this is not the case. We consider for $p, p' > 0$ the commutator of $\rho_{\rm R}(-p)$ and $\rho_{\rm R}(p')$.

$$
\begin{aligned}
[\rho_{\rm R}(-p), \rho_{\rm R}(p')] &= \frac{1}{L}\left[\sum_k R^\dagger(k+p)R(k), \sum_{k'} R^\dagger(k')R(k'+p')\right] \\
&= \frac{1}{L}\sum_{k,k'}\{\delta_{kk'}R^\dagger(k+p)R(k'+p') - \delta_{k+p,k'+p'}R^\dagger(k')R(k)\} \\
&= \frac{1}{L}\sum_k \{R^\dagger(k+p)R(k+p') - R^\dagger(k+p-p')R(k)\}\ .
\end{aligned}
\tag{2.1.16}
$$

The right-hand side is again an operator. We split the expression using normal ordering:

$$
R^\dagger(k+p)R(k+p') = \,:R^\dagger(k+p)R(k+p'):\, + \delta_{p,p'}n^{\rm R}_{k+p}\,,
$$

with $n^{\rm R}_k = \langle F|R^\dagger(k)R(k)|F\rangle$. Then, the second term cancels with the normal product after performing the substitution $k \to k+p'$. The result is

$$
[\rho_{\rm R}(-p), \rho_{\rm R}(p')] = \delta_{pp'}\frac{1}{L}\sum_k \left[n^{\rm R}_{k+p} - n^{\rm R}_k\right]\ . \tag{2.1.17}
$$

Here, because of

$$
n^{\rm R}_k = \begin{cases} 0 & (k > 0) \\ 1 & (k < 0) \end{cases} \tag{2.1.18}
$$

(2.1.17) can be estimated to be

$$
[\rho_{\rm R}(-p), \rho_{\rm R}(p')] = -\frac{p}{2\pi}\delta_{pp'}\ . \tag{2.1.19a}
$$

In the same manner, we obtain

$$
[\rho_{\rm L}(-p), \rho_{\rm L}(p')] = \frac{p}{2\pi}\delta_{pp'}\ . \tag{2.1.19b}
$$

The expressions in the real space representation read

$$
\begin{aligned}
[\rho_{\mathrm{R}}(x), \rho_{\mathrm{R}}(x')] &= \frac{1}{L}\sum_{p,p'} \mathrm{e}^{-\mathrm{i}px+\mathrm{i}p'x'}\,[\rho_{\mathrm{R}}(-p), \rho_{\mathrm{R}}(p')] \\
&= \frac{1}{L}\sum_{p>0} \mathrm{e}^{-\mathrm{i}p(x-x')}\left(-\frac{p}{2\pi}\right) + \frac{1}{L}\sum_{p>0} \mathrm{e}^{\mathrm{i}p(x-x')}\frac{p}{2\pi} \\
&= \frac{1}{L}\sum_{p}\left(-\frac{p}{2\pi}\right)\mathrm{e}^{-\mathrm{i}p(x-x')} = -\frac{\mathrm{i}}{2\pi}\partial_x\delta(x-x')
\end{aligned}
\tag{2.1.20a}
$$

and

$$
[\rho_{\mathrm{L}}(x), \rho_{\mathrm{L}}(x')] = \frac{\mathrm{i}}{2\pi}\partial_x\delta(x-x'). \tag{2.1.20b}
$$

Notice that the derivative in x appears here.

Now, we introduce the phase $\phi_{\mathrm{R,L}}(x)$ defined as the integral of $\rho_{\mathrm{R,L}}(x)$

$$
\rho_{\mathrm{R}}(x) = \partial_x\phi_{\mathrm{R}}(x)/2\pi\,, \tag{2.1.21a}
$$

$$
\rho_{\mathrm{L}}(x) = \partial_x\phi_{\mathrm{L}}(x)/2\pi\,. \tag{2.1.21b}
$$

More explicitly, we write

$$
\begin{aligned}
\phi_{\mathrm{R}}(x) &= \frac{2\pi}{\sqrt{L}}\sum_{p>0}\mathrm{e}^{-\alpha p/2}\frac{1}{\mathrm{i}p}\left\{\mathrm{e}^{\mathrm{i}px}\rho_{\mathrm{R}}(p) - \mathrm{e}^{-\mathrm{i}px}\rho_{\mathrm{R}}(-p)\right\}\,, \\
&\equiv \phi_{\mathrm{R}}^{-}(x) + \phi_{\mathrm{R}}^{+}(x)
\end{aligned}
\tag{2.1.22a}
$$

$$
\begin{aligned}
\phi_{\mathrm{L}}(x) &= \frac{2\pi}{\sqrt{L}}\sum_{p>0}\mathrm{e}^{-\alpha p/2}\frac{1}{\mathrm{i}p}\left\{\mathrm{e}^{\mathrm{i}px}\rho_{\mathrm{L}}(p) - \mathrm{e}^{-\mathrm{i}px}\rho_{\mathrm{L}}(-p)\right\} \\
&\equiv \phi_{\mathrm{L}}^{+}(x) + \phi_{\mathrm{L}}^{-}(x)\,.
\end{aligned}
\tag{2.1.22b}
$$

Here, α is a cut off with magnitude $\alpha \sim 1$.

Now, we define the vacuum state $|0\rangle$ of the bosons ρ_{R} and ρ_{L} corresponding to the ground state $|F\rangle$ of the fermionic system. Notice that the sign at the right-hand side of the commutator (2.1.19) has been fixed by setting $p > 0$. In what follows, ρ_{R} and ρ_{L} are expressed as the bosonic creation and annihilation operators $b^\dagger$ and b. Using (2.1.19) with $p > 0$, we obtain the relations

$$
\rho_{\mathrm{R}}(p) = \sqrt{\frac{p}{2\pi}}\,b_{1,p}\,, \qquad \rho_{\mathrm{R}}(-p) = \sqrt{\frac{p}{2\pi}}\,b_{1,p}^\dagger\,, \tag{2.1.23a}
$$

$$
\rho_{\mathrm{L}}(p) = \sqrt{\frac{p}{2\pi}}\,b_{2,-p}^\dagger\,, \qquad \rho_{\mathrm{L}}(-p) = \sqrt{\frac{p}{2\pi}}\,b_{2,-p}\,. \tag{2.1.23b}
$$

Then, the relation $[b_{i,p}, b_{j,p'}^\dagger] = \delta_{i,j}\delta_{p,p'}$ holds. The vacuum $|0\rangle$ is defined as

$$
\phi_{\mathrm{R}}^{-}(x)|0\rangle = \phi_{\mathrm{L}}^{-}(x)|0\rangle = 0\,.
$$

Using ϕ_R and ϕ_L, we define

$$\theta_+(x) = \phi_\mathrm{R}(x) + \phi_\mathrm{L}(x)\,, \tag{2.1.24a}$$

$$\theta_-(x) = \phi_\mathrm{R}(x) - \phi_\mathrm{L}(x)\,. \tag{2.1.24b}$$

Then, (2.1.14) can be expressed as

$$\rho(x) = \partial_x\theta_+(x)/2\pi\,, \tag{2.1.25a}$$

$$j(x) = \partial_x\theta_-(x)/2\pi\,. \tag{2.1.25b}$$

Next, we construct the commutation relations. The result is

$$\begin{aligned}[\theta_\pm(x),\theta_\pm(x')] &= [\phi_\mathrm{R}(x),\phi_\mathrm{R}(x')] + [\phi_\mathrm{L}(x),\phi_\mathrm{L}(x')] \\ &= 0\,,\end{aligned} \tag{2.1.26a}$$

$$\begin{aligned}[\theta_+(x),\theta_-(x')] &= [\phi_\mathrm{R}(x),\phi_\mathrm{R}(x')] - [\phi_\mathrm{L}(x),\phi_\mathrm{L}(x')] \\ &= +2\pi\mathrm{i}\,\mathrm{sgn}(x-x')\,.\end{aligned} \tag{2.1.26b}$$

Here, we have used that from (2.1.22a) we obtain

$$\begin{aligned}-[\phi_\mathrm{R}(x),\phi_\mathrm{R}(x')] &= \sum_{p>0} \mathrm{e}^{-\alpha p}\frac{2\pi}{pL}\left(-\mathrm{e}^{\mathrm{i}p(x-x')} + \mathrm{e}^{-\mathrm{i}p(x-x')}\right) \\ &= -2\mathrm{i}\int_0^\infty \mathrm{d}p\,\frac{\mathrm{e}^{-\alpha p}\sin p(x-x')}{p}\,.\end{aligned} \tag{2.1.27}$$

This integral converges also for $\alpha \to 0$ and becomes then $-\mathrm{i}\pi\,\mathrm{sgn}(x-x')$. We conclude that θ_+ and θ_- are canonical conjugates to each other. Defining

$$\pi(x) = -\frac{1}{4\pi}\partial_x\theta_-(x)\,, \tag{2.1.28}$$

we obtain from (2.1.26b)

$$[\theta_+(x),\pi(x')] = \mathrm{i}\delta(x-x')\,, \tag{2.1.29}$$

and therefore π is the conjugated momentum of θ_+.

Having defined $\theta_\pm$ and $\phi_\mathrm{R,L}$ in this manner, the fermionic operators can be expressed in these variables. The expressions are

$$R(x) = \frac{1}{\sqrt{2\pi\alpha}}\eta_1\,\mathrm{e}^{\mathrm{i}\phi_\mathrm{R}(x)}\,, \tag{2.1.30a}$$

$$L(x) = \frac{1}{\sqrt{2\pi\alpha}}\eta_2\,\mathrm{e}^{-\mathrm{i}\phi_\mathrm{L}(x)}\,. \tag{2.1.30b}$$

Here, η_1 and η_2 are the so-called Majorana–fermions creating and annihilating real fermions ($\eta_i^\dagger = \eta_i$) and fulfilling the anti-commutation relation $\{\eta_i,\eta_j\} = 2\delta_{ij}$. This operator has not been included in the discussion so far and is related to the 'zero mode' with wave vector $p = 0$, and corresponds to the exponential of Q and P in equation (2.2.79) of the next section. The presence of this Majorana fermion causes the anti-commutation relations between $R^\dagger, R$ and $L^\dagger, L$. In what follows, we will not give an exact proof, but we will just mention important steps with regard to (2.1.30).

I Reproducing the Expression for $\langle F|R^\dagger(x)R(x)|F\rangle$

In terms of the fermionic operators, we obtain the expression

$$\begin{aligned}\langle F|R^\dagger(x)R(x')|F\rangle &= \frac{1}{L}\sum_k \mathrm{e}^{-\mathrm{i}k(x-x')}\langle F|R^\dagger(k)R(k)|F\rangle \\ &= \frac{1}{L}\sum_k \mathrm{e}^{-\mathrm{i}k(x-x')}n_k^{\mathrm{R}} = \frac{1}{L}\sum_{k<0}\mathrm{e}^{-\mathrm{i}k(x-x')} \\ &= \frac{1}{2\pi}\int_{-\infty}^{0}\mathrm{d}k\,\mathrm{e}^{[\varepsilon-\mathrm{i}(x-x')]k} \\ &= \frac{1}{2\pi}\frac{1}{\varepsilon-\mathrm{i}(x-x')}\,. \end{aligned} \tag{2.1.31}$$

Here, we introduced the infinitesimal cut off $\varepsilon > 0$. On the other hand, in terms of the bosons, (2.1.30a) leads to

$$\langle 0|R^\dagger(x)R(x')|0\rangle = \frac{1}{2\pi\alpha}\langle 0|\,\mathrm{e}^{-\mathrm{i}\phi_{\mathrm{R}}(x)}\,\mathrm{e}^{\mathrm{i}\phi_R(x')}|0\rangle\,. \tag{2.1.32}$$

Using the following formula

$$\mathrm{e}^{A+B} = \mathrm{e}^A\mathrm{e}^B\mathrm{e}^{-\frac{1}{2}[A,B]} = \mathrm{e}^B\mathrm{e}^A\mathrm{e}^{\frac{1}{2}[A,B]}\,, \tag{2.1.33}$$

being valid in the case that $[A,B]$ is a c-number, we obtain from (2.1.22a)

$$\begin{aligned}&\exp\{-\mathrm{i}\phi_{\mathrm{R}}(x)\}\exp\{\mathrm{i}\phi_{\mathrm{R}}(x')\} \\ &= \exp\{\mathrm{i}(-\phi_{\mathrm{R}}(x)+\phi_{\mathrm{R}}(x'))\}\exp\{\tfrac{1}{2}[\phi_{\mathrm{R}}(x),\phi_{\mathrm{R}}(x')]\} \\ &= \exp\{\mathrm{i}(-\phi^+_{\mathrm{R}}(x)+\phi^+_{\mathrm{R}}(x'))+\mathrm{i}(-\phi^-_{\mathrm{R}}(x)+\phi^-_{\mathrm{R}}(x'))\}\exp\{\tfrac{1}{2}[\phi_{\mathrm{R}}(x),\phi_{\mathrm{R}}(x')]\} \\ &= \exp\{\mathrm{i}(-\phi^+_{\mathrm{R}}(x)+\phi^+_{\mathrm{R}}(x'))\}\exp\{\mathrm{i}(-\phi^-_{\mathrm{R}}(x)+\phi^-_{\mathrm{R}}(x'))\} \\ &\quad\times\exp\{\tfrac{1}{2}[-\phi^+_{\mathrm{R}}(x)+\phi^+_{\mathrm{R}}(x'),-\phi^-_{\mathrm{R}}(x)+\phi^-_{\mathrm{R}}(x')]+\tfrac{1}{2}[\phi_{\mathrm{R}}(x),\phi_{\mathrm{R}}(x')]\}\end{aligned} \tag{2.1.34}$$

and using

$$\langle 0|\exp\{\mathrm{i}(-\phi^+_{\mathrm{R}}(x)+\phi^+_{\mathrm{R}}(x'))\} = \langle 0|\,,\qquad \exp\{\mathrm{i}(-\phi^-_{\mathrm{R}}(x)+\phi^-_{\mathrm{R}}(x'))\}|0\rangle = |0\rangle\,,$$

we obtain

$$\langle 0|R^\dagger(x)R(x')|0\rangle = \frac{1}{2\pi\alpha}\exp\left\{[\phi^+_R(0),\ \phi^-_R(0)] - [\phi^+_R(x'),\ \phi^-_R(x)]\right\}. \tag{2.1.35}$$

From equation (2.1.22a) and the commutator (2.1.19a), we obtain

$$\langle 0|R^\dagger(x)R(x')|0\rangle = \frac{1}{2\pi\alpha}\exp\left[-\frac{2\pi}{L}\sum_{p>0}\mathrm{e}^{-\alpha p}\frac{1-\mathrm{e}^{\mathrm{i}p(x-x')}}{p}\right]\,, \tag{2.1.36}$$

and for $L\to\infty$, the sum in the exponential becomes an integral:

$$\int_0^\infty \mathrm{d}p\,\mathrm{e}^{-\alpha p}\frac{1-\mathrm{e}^{\mathrm{i}p(x-x')}}{p} = -\int_0^\infty \mathrm{d}p\,\mathrm{e}^{-\alpha p}\sum_{n=1}^{\infty}\frac{\mathrm{i}^n}{n!}(x-x')^n p^{n-1}$$
$$= -\sum_{n=1}^{\infty}\frac{\mathrm{i}^n}{n!}(x-x')^n\frac{(n-1)!}{\alpha^n}$$
$$= -\sum_{n=1}^{\infty}\frac{1}{n}\left[\frac{\mathrm{i}(x-x')}{\alpha}\right]^n = \ln\left[1-\frac{\mathrm{i}(x-x')}{\alpha}\right], \quad (2.1.37)$$

and finally, we obtain

$$\langle 0|R^\dagger(x)R(x')|0\rangle = \frac{1}{2\pi}\frac{1}{\alpha-\mathrm{i}(x-x')}. \quad (2.1.38)$$

With the correspondence $\alpha \leftrightarrow \varepsilon$, this expression agrees with (2.1.31).

II Determination of the Anti-Commutation Relation

From (2.1.30) and (2.1.31), we obtain

$$R(x)R(x') = \frac{1}{2\pi\alpha}\,\mathrm{e}^{\mathrm{i}\phi_\mathrm{R}(x)}\,\mathrm{e}^{\mathrm{i}\phi_\mathrm{R}(x')}$$
$$= \frac{1}{2\pi\alpha}\,\mathrm{e}^{\mathrm{i}\phi_\mathrm{R}(x')}\,\mathrm{e}^{\mathrm{i}\phi_\mathrm{R}(x)}\,\mathrm{e}^{[\phi_\mathrm{R}(x),\phi_\mathrm{R}(x')]} \quad (2.1.39)$$

and from $[\phi_\mathrm{R}(x),\phi_\mathrm{R}(x')] = \mathrm{i}\pi\,\mathrm{sgn}(x-x')$ obviously $\exp\{[\phi_\mathrm{R}(x),\phi_\mathrm{R}(x')]\} = -1$. Therefore, we obtain $R(x)R(x') = -R(x')R(x)$ from (2.1.39).

III The Commutation Relation of the Fermion Operator and the Density Operator

In terms of the fermions, we obtain

$$[R(x),\rho_\mathrm{R}(p)] = \left[\frac{1}{\sqrt{L}}\sum_k \mathrm{e}^{\mathrm{i}kx}R(k), \frac{1}{\sqrt{L}}\sum_{k'} R^\dagger(k')R(k'+p)\right]$$
$$= \frac{1}{L}\sum_k \mathrm{e}^{\mathrm{i}kx}R(k+p) = \frac{1}{\sqrt{L}}\mathrm{e}^{-\mathrm{i}px}R(x), \quad (2.1.40)$$

and in terms of the bosons

$$[R(x),\rho_\mathrm{R}(p)] = \frac{1}{\sqrt{2\pi\alpha}}\eta_1\{\mathrm{e}^{\mathrm{i}\phi_\mathrm{R}(x)}\rho_\mathrm{R}(p) - \rho_\mathrm{R}(p)\,\mathrm{e}^{\mathrm{i}\phi_\mathrm{R}(x)}\}$$
$$= \{\mathrm{e}^{\mathrm{i}\phi_\mathrm{R}(x)}\rho_\mathrm{R}(p)\,\mathrm{e}^{-\mathrm{i}\phi_\mathrm{R}(x)} - \rho_\mathrm{R}(p)\}R(x)$$
$$= \int_0^1 \mathrm{d}\eta\,\frac{\mathrm{d}}{\mathrm{d}\eta}\left[\mathrm{e}^{\mathrm{i}\eta\phi_\mathrm{R}(x)}\rho_\mathrm{R}(p)\,\mathrm{e}^{-\mathrm{i}\eta\phi_\mathrm{R}(x)}\right]R(x)$$

$$= \int_0^1 d\eta\, e^{i\eta\phi_R(x)}\, [i\phi_R(x), \rho_R(p)]\, e^{-i\eta\phi_R(x)} R(x)$$
$$= \frac{1}{\sqrt{L}} e^{-ipx-\alpha p/2} R(x)\,. \tag{2.1.41}$$

For $\alpha \to 0$, these expressions do agree.

So, what might be the physical meaning of (2.1.29)? Using $\phi_R(x) = \frac{1}{2}(\theta_+(x) + \theta_-(x))$, $\phi_L(x) = \frac{1}{2}(\theta_+(x) - \theta_-(x))$, we can write

$$R(x) = \frac{1}{\sqrt{2\pi\alpha}} \eta_1\, e^{i\frac{\theta_+(x)+\theta_-(x)}{2}}\,, \tag{2.1.42a}$$

$$L(x) = \frac{1}{\sqrt{2\pi\alpha}} \eta_2\, e^{i\frac{-\theta_+(x)+\theta_-(x)}{2}}\,. \tag{2.1.42b}$$

Combining this equation with (2.1.26), the significance of $\theta_\pm$ as well as R and L can be seen. First,

$$R^\dagger(x)L(x) = \frac{1}{2\pi\alpha}\eta_1\eta_2\, e^{-i\theta_+(x)} \tag{2.1.43}$$

is the term appearing in the alternating density term of (2.1.11) in the vicinity of the wave number $2k_F = \pi$. That is, we obtain

$$\rho_{2k_F}(x) = \frac{1}{\pi\alpha}\cos(\theta_+(x) - 2k_F x)\,, \tag{2.1.44}$$

where θ_+ is the phase of the particle density wave. When the phase θ_+ varies spatially, the compression and expansion of the density wave occurs. This results in the density fluctuation as expressed in (2.1.25a)

On the other hand, a Cooper pair in the vicinity of the wave number zero is given by the product of fermionic operators with wave number k_F and $-k_F$

$$R(x)L(x) = \frac{1}{2\pi\alpha}\eta_1\eta_2\, e^{i\theta_-(x)}\,, \tag{2.1.45}$$

and therefore we identify $\theta_-(x)$ as the Josephson phase of a superconductor. Then, (2.1.25b) is nothing but the Josephson equation. Furthermore, the canonical commutation relation between θ_+ and θ_- can be interpreted as the one-dimensional version of that between particle number and phase.

Next, what might be the interpretation of $R(x)$ and $L(x)$ themselves? In the previous chapter, we saw that the fermionic operator corresponds to the kink operator, and this property is preserved also after the bosonization. Now, we consider a state $|\theta_0\rangle$ with $\theta_+(x) = \theta_0 = \text{const}$ and act on it with $R(x)$. $R^\dagger(x)$ containing $e^{-\frac{i}{2}\theta_-(x)}$ acts on $\theta_+(x)$ as the unitary linear translation operator. Defining

$$U_\eta(x) \equiv e^{i\eta\frac{\theta_-(x)}{2}}\,, \tag{2.1.46}$$

and taking the derivative of

$$\theta_+(x', \eta) \equiv U_\eta^\dagger(x)\theta_+(x')U_\eta(x)\,, \tag{2.1.47}$$

we obtain

$$\begin{aligned}\frac{\partial\theta_+(x',\eta)}{\partial\eta} &= U_\eta^\dagger(x)\left[-\mathrm{i}\frac{\theta_-(x)}{2}, \theta_+(x')\right]U_\eta(x) \\ &= \pi\,\mathrm{sgn}(x-x')\,,\end{aligned} \tag{2.1.48}$$

and therefore

$$\theta_+(x', \eta = 1) = \theta_+(x') + \pi\,\mathrm{sgn}(x - x')\,. \tag{2.1.49}$$

We conclude that a kink in θ_+ at $x = x'$ has been created, where a 2π shift occurs. Notice the intimate relation between the kink creation and the non-locality of the commutator (2.1.26b).

After these preliminaries, we want to express (2.1.7) and the J_z term in terms of the bosons. First, in order to bosonize H_{XY}, we rewrite the Hamiltonian as $\tilde{H}_{\mathrm{B}}$ which is quadratic in the boson operators satisfying the same commutation relation as H_{XY} with $\rho_{\mathrm{R,L}}(p)$. First, we consider the fermionic Hamiltonian and obtain

$$\begin{aligned}[\rho_{\mathrm{R}}(p), H_{XY}] &= \left[\sum_k R^\dagger(k)R(k+p), \sum_{k'} J_\perp k' R^\dagger(k')R(k')\right] \\ &= J_\perp \sum_{k,k'} k' \left\{R^\dagger(k)R(k')\delta_{k+p,k'} - R^\dagger(k')R(k+p)\delta_{kk'}\right\} \\ &= J_\perp \sum_k \left\{(k+p)R^\dagger(k)R(k+p) - kR^\dagger(k)R(k+p)\right\} \\ &= J_\perp p\rho_{\mathrm{R}}(p)\,,\end{aligned} \tag{2.1.50}$$

and

$$[\rho_{\mathrm{L}}(p), H_{XY}] = -J_\perp p\rho_{\mathrm{L}}(p)\,. \tag{2.1.51}$$

The same commutation relations are satisfied by $\tilde{H}_{\mathrm{B}}$ given by

$$\tilde{H}_{\mathrm{B}} = 2\pi J_\perp \sum_{p>0} \left\{\rho_{\mathrm{R}}(-p)\rho_{\mathrm{R}}(p) + \rho_{\mathrm{L}}(p)\rho_{\mathrm{L}}(-p)\right\}\,. \tag{2.1.52}$$

Indeed, we obtain again

$$\begin{aligned}[\rho_{\mathrm{R}}(p), \tilde{H}_{\mathrm{B}}] &= 2\pi J_\perp [\rho_{\mathrm{R}}(p), \rho_{\mathrm{R}}(-p)\rho_{\mathrm{R}}(p)] \\ &= 2\pi J_\perp \frac{p}{2\pi}\rho_{\mathrm{R}}(p) = J_\perp p\rho_{\mathrm{R}}(p)\,.\end{aligned} \tag{2.1.53}$$

In the coordinate space, up to constant terms, we obtain

$$\begin{aligned}\tilde{H}_{\mathrm{B}} &= \pi J_\perp \int \mathrm{d}x \left\{\rho_{\mathrm{R}}(x)^2 + \rho_{\mathrm{L}}(x)^2\right\} \\ &= \frac{J_\perp}{8\pi}\int \mathrm{d}x \left\{(\partial_x\theta_+(x))^2 + (\partial_x\theta_-(x))^2\right\}\,.\end{aligned} \tag{2.1.54}$$

Next, for the interaction part between the fermions proportional to J_z, we obtain from (2.1.11) and (2.1.44)

$$H_{J_z} = J_z \int \mathrm{d}x \left\{ (\rho_{\mathrm{R}}(x) + \rho_{\mathrm{L}}(x))^2 - \left(\frac{1}{2\pi\alpha} \left(\mathrm{e}^{\mathrm{i}(\phi_{\mathrm{R}}+\phi_{\mathrm{L}})} + \mathrm{e}^{-\mathrm{i}(\phi_{\mathrm{R}}+\phi_{\mathrm{L}})} \right) \right)^2 \right\}$$

$$= J_z \int \mathrm{d}x \left\{ \left(\frac{1}{2\pi} \partial_x \theta_+(x) \right)^2 - \left(\frac{1}{\pi\alpha} \cos\theta_+(x) \right)^2 \right\} . \qquad (2.1.55)$$

Using $\cos^2\theta_+ = \frac{1}{2}(1+\cos 2\theta_+)$, finally the Hamiltonian becomes

$$H = \int \mathrm{d}x \left\{ \left(\frac{J_\perp}{8\pi} + \frac{J_z}{4\pi^2} \right) (\partial_x\theta_+)^2 + \frac{J_\perp}{8\pi} (\partial_x\theta_-)^2 - \frac{J_z}{2(\pi\alpha)^2} \cos 2\theta_+ \right\} . \qquad (2.1.56)$$

This is nothing more than the quantum Sine–Gordon model. Using the momentum $\pi(x)$ introduced in (2.1.28), (2.1.56) can be expressed as

$$H = \int \mathrm{d}x \left\{ A\pi(x)^2 + C(\partial_x\theta_+(x))^2 - B\cos 2\theta_+(x) \right\} . \qquad (2.1.57)$$

Here, we introduced the constants

$$A = 2\pi J_\perp , \qquad (2.1.58a)$$

$$C = \frac{1}{8\pi} \left(J_\perp + \frac{2J_z}{\pi} \right) , \qquad (2.1.58b)$$

$$B = \frac{J_z}{2(\pi\alpha)^2} . \qquad (2.1.58c)$$

First, we consider the Hamiltonian H_0 with B set to zero. This is the Hamiltonian of the harmonic oscillator with a linear dispersion relation. In (2.1.57), A and C are the coefficients endeavouring to fix the momentum π and the coordinate θ_+, respectively, and their ratio should determine the dynamics of the canonical pair. That is, when A/C is large, π or θ_- is fixed, in the interpretation of the spin model, this corresponds to the singlet formation, and, in terms of the fermion model, the superconductive correlation dominates. On the other hand, when A/C is small, θ_+ is fixed, the spins are anti-ferromagnetically long-range ordered, and in the fermionic model the charge density wave dominates. Now, we introduce the quantum parameter η with the following formula:

$$\eta \equiv \frac{1}{2\pi} \sqrt{\frac{A}{C}} . \qquad (2.1.59)$$

Another important quantity having the dimension of an energy is given by

$$v = 2\sqrt{AC} . \qquad (2.1.60)$$

As will be explained below, this is the sound velocity for the linear dispersion relation. With (2.1.29), the Heisenberg equations of motion are given by

$$\begin{aligned}\frac{\partial\theta_+(x,t)}{\partial t} &= \frac{1}{\mathrm{i}}[\theta_+(x,t),H_0] \\ &= \frac{1}{\mathrm{i}}\int \mathrm{d}x'\, 2A[\theta_+(x,t),\pi(x',t)]\pi(x',t) \\ &= 2A\pi(x,t)\,, \qquad (2.1.61\mathrm{a})\end{aligned}$$

$$\begin{aligned}\frac{\partial\pi(x,t)}{\partial t} &= \frac{1}{\mathrm{i}}[\pi(x,t),H_0] \\ &= \frac{1}{\mathrm{i}}\int \mathrm{d}x'\, 2C\partial_{x'}\theta_+(x',t)[\pi(x,t),\partial_{x'}\theta_+(x',t)] \\ &= \frac{1}{\mathrm{i}}\int \mathrm{d}x'\, 2C\partial_{x'}\theta_+(x',t)(-\mathrm{i})\partial_{x'}\delta(x'-x) \\ &= 2C\partial_x^2\theta_+(x,t)\,. \qquad (2.1.61\mathrm{b})\end{aligned}$$

Therefore, we obtain for θ_+ the wave equation

$$\frac{\partial^2\theta_+(x,t)}{\partial t^2} = 4AC\frac{\partial^2\theta_+(x,t)}{\partial x^2}\,. \qquad (2.1.62)$$

We conclude that the dispersion relation is given by

$$\omega^2 = 4ACq^2 = v^2q^2\,. \qquad (2.1.63)$$

Because A and B are given in terms of $J_\perp$ and J_z by (2.1.58), v (2.1.60) and η (2.1.59) can be determined. However, because (2.1.58) has been derived in the continuum limit, it is only valid for $|J_z| \ll J_\perp$, and we cannot trust it any more for $|J_z| \sim J_\perp$. What about the Hamiltonian (2.1.57) for $|J_z| \sim J_\perp$? When restricted to long wavelengths (small frequencies), (2.1.57) is still the right description for the physics. When no gap in the spectrum arises (that is, for $|J_z| \le |J_\perp|$), the low-energy modes are correctly described by (2.1.57). Then, the question arises as to how η and v should be chosen in general. Below, we examine the case of the Heisenberg model ($J_z = J_\perp = J$).

First, v should have the property that the linear dispersion of the des Cloiseaux-Pearson mode derived in Sect. 1.3 is reproduced:

$$v = \frac{\pi}{2}J\,. \qquad (2.1.64)$$

On the other hand, as will become clear in what follows, η can be determined because it is the critical exponent determining the behaviour of some correlation functions. To do so, we rewrite the Hamiltonian using v and η as

$$\begin{aligned}H_0 &= v\int \mathrm{d}x \left\{\pi\eta\pi(x)^2 + \frac{1}{4\pi\eta}(\partial_x\theta_+(x))^2\right\} \\ &= \int \mathrm{d}x\, \frac{v}{4\pi}\left\{\frac{\eta}{4}(\partial_x\theta_-(x))^2 + \frac{1}{\eta}(\partial_x\theta_+(x))^2\right\}\,. \qquad (2.1.65)\end{aligned}$$

The corresponding Lagrangian is given by

$$\begin{aligned} L &= \int \pi(x)\partial_t\theta_+(x)\,\mathrm{d}x - H_0 \\ &= \int \mathrm{d}x \left\{ -\frac{1}{4\pi}\partial_x\theta_-(x)\partial_t\theta_+(x) - \frac{v}{4\pi}\left[\frac{\eta}{4}(\partial_x\theta_-(x))^2 + \frac{1}{\eta}(\partial_x\theta_+(x))^2\right]\right\}. \end{aligned} \tag{2.1.66}$$

Integrating out θ_-, we obtain

$$L_{\theta_+} = \int \mathrm{d}x\,\frac{1}{4\pi\eta}\left[\frac{1}{v}(\partial_t\theta_+(x))^2 - v(\partial_x\theta_+(x))^2\right]. \tag{2.1.67}$$

On the other hand, integrating out θ_+, we obtain

$$L_{\theta_-} = \int \mathrm{d}x\,\frac{\eta}{16\pi}\left[\frac{1}{v}(\partial_t\theta_-(x))^2 - v(\partial_x\theta_-(x))^2\right]. \tag{2.1.68}$$

On the basis of the above discussion, we deduce a representation of the spin operators in terms of θ_+ and θ_-. First, using (1.2.1c) and (2.1.11), S_n^z can be expressed as

$$S_n^z = \frac{1}{2\pi}\partial_x\theta_+(x_n) + (-1)^n\frac{1}{\pi\alpha}\cos\theta_+(x_n). \tag{2.1.69}$$

The relationship between $S_n^\pm$ and the fermions is given in (1.2.1a,b). Now, it becomes necessary to consider the non-local operator $K(n)$:

$$\begin{aligned} K(n) &= \exp\left[\mathrm{i}\pi\sum_{j=1}^{n-1} f_j^\dagger f_j\right] = \exp\left[\frac{\mathrm{i}\pi(n-1)}{2} + \mathrm{i}\pi\sum_{j=1}^{n-1} :f_j^\dagger f_j:\right] \\ &\cong \exp\left[\frac{\mathrm{i}\pi(n-1)}{2} + \mathrm{i}\pi\int_0^{x_n}\frac{1}{2\pi}\partial_x\theta_+(x')\,\mathrm{d}x'\right] \\ &= \exp\left[\frac{\mathrm{i}\pi(n-1)}{2} + \frac{\mathrm{i}}{2}\left[\theta_+(x) - \theta_+(0)\right]\right] \\ &= \text{const} \times \exp\left\{\mathrm{i}\frac{\pi n}{2} + \mathrm{i}\frac{\theta_+(x)}{2}\right\}. \end{aligned} \tag{2.1.70}$$

We obtain

$$\begin{aligned} S_n^+ &\cong \left[R^\dagger(x_n)\exp\left[-\mathrm{i}\frac{\pi}{2}n\right] + L^\dagger(x_n)\exp\left[\mathrm{i}\frac{\pi}{2}n\right]\right]\exp\left[\mathrm{i}\frac{\pi}{2}n + \mathrm{i}\frac{\theta_+(x)}{2}\right] \\ &= \frac{1}{\sqrt{2\pi\alpha}}\left[\exp\left[-\mathrm{i}\frac{\theta_+(x)+\theta_-(x)}{2} + \mathrm{i}\frac{\theta_+(x)}{2}\right]\right. \\ &\qquad \left. +(-1)^n\exp\left[\mathrm{i}\frac{\theta_+(x)-\theta_-(x)}{2} + \mathrm{i}\frac{\theta_+(x)}{2}\right]\right] \\ &= \frac{1}{\sqrt{2\pi\alpha}}\left[\exp\left[-\mathrm{i}\frac{\theta_-(x)}{2}\right] + (-1)^n\exp\left[\mathrm{i}\theta_+(x) - \frac{1}{2}\theta_-(x)\right]\right]. \end{aligned} \tag{2.1.71}$$

Keeping in mind the sign convention made in (1.2.11) (that is, $J_\perp \to -J_\perp$), the alternating component can be expressed as

$$(S_n^z)_{\text{altern.}} = M_n^z \sim \cos\theta_+(x_n)\,, \tag{2.1.72a}$$

$$(S_n^x)_{\text{altern.}} = M_n^x \sim \cos\frac{\theta_-(x_n)}{2}\,. \tag{2.1.72b}$$

To deal with space and time components in a symmetric way, we will now use the imaginary time formalism. Notice that the action

$$A := \int \mathrm{d}t\, L$$

appears in the path integral as

$$\mathrm{e}^{\mathrm{i}A}\,.$$

Passing to the imaginary time formalism by setting $\tau = it$, because the above equation changes to

$$\mathrm{e}^{-A}\,,$$

the action becomes

$$A = -\int_0^\beta \mathrm{d}\tau \int \mathrm{d}x\, L(\tau = \mathrm{i}t)\,.$$

Here, $\beta = 1/T$ is the inverse of the temperature. Explicitly, for example (2.1.67) becomes

$$A_0 = \int_0^\beta \mathrm{d}\tau \int \mathrm{d}x\, \frac{1}{4\pi\eta}\left[\frac{1}{v}(\partial_\tau\theta_+(x,\tau))^2 + v(\partial_x\theta_+(x,\tau))^2\right]\,. \tag{2.1.73}$$

We now calculate the correlation function of $M^z(x,\tau)$ for this action. With T_τ being the time-ordered product, we obtain

$$\langle T_\tau\, \mathrm{e}^{\mathrm{i}\theta_+(x,\tau)}\, \mathrm{e}^{-\mathrm{i}\theta_+(0,0)}\rangle = \frac{1}{Z}\int \mathcal{D}\theta_+(x,\tau)\, \exp\{\mathrm{i}(\theta_+(x,\tau) - \theta_+(0,0)) - S_0\}\,. \tag{2.1.74}$$

Here, the partition function is given by

$$Z = \int \mathcal{D}\theta_+(x,\tau)\,\mathrm{e}^{-S_0}\,,$$

(2.1.73) is quadratic in θ_+, and introducing the Fourier transformation

$$\theta_+(x,\tau) = \frac{1}{\sqrt{\beta L}}\sum_{\mathrm{i}\omega_n}\sum_q \mathrm{e}^{\mathrm{i}qx - \mathrm{i}\omega_n\tau}\theta_+(q,\mathrm{i}\omega_n) \qquad (\omega_n = 2\pi T n)\,, \tag{2.1.75}$$

we can write (2.1.73) as

$$A_0 = \sum_{\mathrm{i}\omega_n} \sum_q \frac{\omega_n^2 + v^2 q^2}{4\pi\eta v} \theta_+(q, \mathrm{i}\omega_n)\theta_+(-q, -\mathrm{i}\omega_n) \,. \tag{2.1.76}$$

Therefore, (2.1.74) becomes

$$\begin{aligned} &\left\langle T_\tau \, \mathrm{e}^{\mathrm{i}\theta_+(x,\tau)} \, \mathrm{e}^{-\mathrm{i}\theta_+(0,0)} \right\rangle \\ &= \left\langle \exp\left[\mathrm{i} \frac{1}{\sqrt{\beta L}} \sum_{q,\mathrm{i}\omega_n} \left(\mathrm{e}^{\mathrm{i}qx - \mathrm{i}\omega_n \tau} - 1 \right) \theta_+(q, \mathrm{i}\omega_n) \right] \right\rangle \\ &= \exp\left[-\frac{1}{\beta L} \sum_{q,\mathrm{i}\omega_n} \frac{\pi\eta v}{\omega_n^2 + v^2 q^2} \left(1 - \mathrm{e}^{\mathrm{i}qx - \mathrm{i}\omega_n \tau}\right) \left(1 - \mathrm{e}^{-\mathrm{i}qx + \mathrm{i}\omega_n \tau}\right) \right] . \end{aligned} \tag{2.1.77}$$

In the limit $\beta \to \infty, L \to \infty$, the sum becomes an integral, and the right-hand side of (2.1.77) can be estimated using

$$\begin{aligned} &\exp\left[-\frac{1}{(2\pi)^2} \int \mathrm{d}\omega \, \mathrm{d}q \, \frac{2\pi\eta v}{\omega^2 + v^2 q^2} \left[1 - \cos(qx - \omega\tau)\right] \right] \\ &\sim \exp\left[-\eta \ln \sqrt{x^2 + v^2\tau^2} \right] = \frac{1}{(x^2 + v^2\tau^2)^{\eta/2}} \,. \end{aligned} \tag{2.1.78}$$

Here, we introduced an infrared cut-off $|\boldsymbol{k}| \sim (x^2+v^2\tau^2)^{-1/2}$ when performing the integral of the two-vector $\boldsymbol{k} = (\omega, vq)$. We conclude that

$$\langle M^z(x,\tau) M^z(0,0) \rangle \sim (x^2 + v^2\tau^2)^{-\eta/2} \,, \tag{2.1.79}$$

and in the same manner, from (2.1.68) and (2.1.72b) we can deduce that

$$\langle M^x(x,\tau) M^x(0,0) \rangle \sim (x^2 + v^2\tau^2)^{-1/2\eta} \,. \tag{2.1.80}$$

Because in the Heisenberg model there is no distinction between the x-component and the z-component, (2.1.79) and (2.1.80) must behave in the same manner. From this requirement, we conclude that $\eta = 1$.

Next, we discuss a term that has been neglected so far, that is, the Umklapp scattering. Expressing this process with fermionic operators, the term is given by $R^\dagger R^\dagger LL + \mathrm{h.c.}$ It induces a momentum transfer of $\pm 4k_\mathrm{F} = \pm 2\pi$. Using (2.1.42), the action containing the Umklapp scattering is given in the imaginary time formalism by

$$\begin{aligned} A = \int_0^\beta \mathrm{d}\tau \int \mathrm{d}x \biggl\{ &\frac{1}{4\pi\eta} \left[\frac{1}{v} (\partial_\tau \theta_+(x,\tau))^2 + v (\partial_x \theta_+(x,\tau))^2 \right] \\ &- \frac{J_z}{2(\pi\alpha)^2} \cos 2\theta_+(x,\tau) \biggr\} \,. \end{aligned} \tag{2.1.81}$$

Here, we will rewrite the action in a slightly different way. First, we change the range of the $\tau-$integration from $\theta_+(x,\tau) = \theta_+(x,\tau+\beta)$ to $(-\beta/2, \beta/2)$, then we consider the limit of zero temperature with an integration region

$(-\infty, +\infty)$. Furthermore, setting $r_0 = v\tau$, $r_1 = x$, we define $\phi(x,\tau)$ by $\theta_+(x,\tau) = \sqrt{2\pi\eta}\phi(x,\tau)$. Then, (2.1.81) becomes

$$A = \int_{-\infty}^{\infty} \mathrm{d}^2 r \left[\frac{1}{2}(\nabla\phi(r))^2 - \mu\cos(\zeta\phi(x,\tau))\right] . \tag{2.1.82}$$

Here, we defined $\xi = 2\sqrt{2\pi\eta}$, $\mu = J_z/(2(\pi\alpha)^2 v)$.

In what follows, we discuss (2.1.82) using the renormalization group method. First, we notice that since the action (2.1.82) is given in the continuum limit, it cannot be used in the short-wavelength limit. Therefore, we introduce a cut-off of the momentum p. As can be understood easily, the cut-off Λ is of the order of the inverse of the lattice constant a of the original lattice model

$$\Lambda \sim a^{-1} \sim \alpha^{-1} .$$

The renormalization group is a method of deducing the effective action of the system in the long-wavelength and low-frequency limit, when the cutoff Λ becomes smaller and smaller. Explicitly, we consider

$$\phi_\Lambda(r) = \int_{0<p<\Lambda} \frac{\mathrm{d}^2 p}{(2\pi)^2}\, \mathrm{e}^{\mathrm{i}pr} \phi(p) . \tag{2.1.83}$$

(Notice $\phi(p)^* = \phi(-p)$). When the cutoff Λ is chosen to be slightly smaller

$$\phi_\Lambda(r) = \phi_{\Lambda'}(r) + h(r) , \tag{2.1.84}$$

$$h(r) = \int_{\Lambda'<p<\Lambda} \frac{\mathrm{d}^2 p}{(2\pi)^2}\, \mathrm{e}^{\mathrm{i}pr} \phi(p) , \tag{2.1.85}$$

then the path integral in $h(r)$ is performed. That is, in

$$Z_\Lambda = \int \mathcal{D}\phi_\Lambda\, \mathrm{e}^{-A(\phi_\Lambda)} = \int \mathcal{D}\phi_{\Lambda'}\mathcal{D}h\, \mathrm{e}^{-A(\phi_{\Lambda'}+h)} \equiv \int \mathcal{D}\phi_{\Lambda'}\, \mathrm{e}^{-A'(\phi_{\Lambda'})} , \tag{2.1.86}$$

$A'(\phi_{\Lambda'})$ is the long-wavelength effective action of $\phi_{\Lambda'}$.

We apply this method explicitly and determine $A'(\phi_{\Lambda'})$.

$$\begin{aligned} \mathrm{e}^{-A'(\phi_{\Lambda'})} &= \int \mathcal{D}h \exp\left[-\int\left\{\frac{1}{2}\left[(\nabla\phi_{\Lambda'}(r))^2 + (\nabla h(r))^2\right]\right.\right. \\ &\quad \left.\left. - \mu\cos\zeta(\phi_{\Lambda'} + h)\right\}\mathrm{d}^2 r\right] \\ &= \text{const} \times \exp\left[-\int \frac{1}{2}(\nabla\phi_{\Lambda'}(r))^2\, \mathrm{d}^2 r\right] \left\langle \mathrm{e}^{\mu\int \cos\zeta(\phi_{\Lambda'}+h)\,\mathrm{d}^2 r}\right\rangle_h . \end{aligned} \tag{2.1.87}$$

Here, we introduced the notation

$$\langle C\rangle_h = \int \mathcal{D}h\, \mathrm{e}^{-\int \frac{1}{2}(\nabla h)^2\, \mathrm{d}^2 r} C \Big/ \int \mathcal{D}h\, \mathrm{e}^{-\int \frac{1}{2}(\nabla h)^2\, \mathrm{d}^2 r} . \tag{2.1.88}$$

Now, we restrict the considerations to the case where μ is small enough. Then, (2.1.87) can be treated perturbatively in μ. However, notice that this is not a straightforward perturbative expansion. That is, when we would all at once integrate in (2.1.82) over the whole ϕ_Λ region and would try to make a perturbative expansion in μ, infrared divergences coming from the small p-region might occur. However, because in (2.1.87) and (2.1.88) only the small-wavelength components of $h(r)$ are integrated out, this problem will not occur.

$$\begin{aligned}
&\left\langle \exp\left[\mu \int \cos\zeta(\phi_{\Lambda'} + h)\, \mathrm{d}^2 r\right]\right\rangle_h \\
&\quad = 1 + \mu \int \langle \cos\zeta(\phi_{\Lambda'} + h)\rangle_h\, \mathrm{d}^2 r \\
&\qquad + \frac{1}{2}\mu^2 \int \mathrm{d}^2 r\, \mathrm{d}^2 r'\ \langle \cos\zeta(\phi_{\Lambda'}(r) + h(r)) \cos\zeta(\phi_{\Lambda'}(r') + h(r'))\rangle_h \,.
\end{aligned} \tag{2.1.89}$$

We consider now the first-order term in μ

$$\begin{aligned}
\langle \cos\zeta(\phi_{\Lambda'}(r) + h(r))\rangle_h &= \tfrac{1}{2}\left(\exp[\mathrm{i}\zeta\phi_{\Lambda'}(r)]\langle \mathrm{e}^{\mathrm{i}\zeta h(r)}\rangle_h + \text{h.c.}\right) \\
&= \exp\left\{-\tfrac{1}{2}\zeta^2 G_h(0)\right\} \cos\zeta\phi_{\Lambda'}(r) \,.
\end{aligned} \tag{2.1.90}$$

Defining the propagator

$$G_h(r) = \int_{\Lambda' < p < \Lambda} \frac{\mathrm{d}^2 p}{(2\pi)^2}\, \mathrm{e}^{\mathrm{i}pr}\, \frac{1}{p^2} \tag{2.1.91}$$

and defining

$$B(r) = \exp\left\{-\tfrac{1}{2}\zeta^2 G_h(r)\right\} , \tag{2.1.92}$$

(2.1.90) becomes

$$\langle \cos\zeta(\phi_{\Lambda'}(r) + h(r))\rangle_h = B(0) \cos\zeta\phi_{\Lambda'}(r) \,. \tag{2.1.93}$$

The term $B(0)$ will be analysed later on.

Next, we consider the second-order term in μ. In the same manner, we obtain

$$\begin{aligned}
&\langle \cos\zeta(\phi_{\Lambda'}(r) + h(r)) \cos\zeta(\phi_{\Lambda'}(r') + h(r'))\rangle_h \\
&\qquad - \langle \cos\zeta(\phi_{\Lambda'}(r) + h(r))\rangle_h \langle \cos\zeta(\phi_{\Lambda'}(r') + h(r'))\rangle_h \\
&\quad = \tfrac{1}{2}B^2(0)\left[B^2(r - r') - 1\right] \cos\zeta(\phi_{\Lambda'}(r) + \phi_{\Lambda'}(r')) \\
&\qquad + \tfrac{1}{2}B^2(0)\left[B^{-2}(r - r') - 1\right] \cos\zeta(\phi_{\Lambda'}(r) - \phi_{\Lambda'}(r')) \,.
\end{aligned} \tag{2.1.94}$$

The cumulant expansion will appear in the exponential function when $A'(\phi_\Lambda)$ is determined later on.

We now discuss the properties of $G_h(r)$ and $B(r)$. Following (2.1.91), because $G_h(r)$ only contains contributions with wave numbers larger than Λ', we expect that for $|r| > \Lambda'^{-1}$ it becomes smaller, and that in this case $B(r) \simeq 1$. For $|r - r'| < \Lambda'^{-1}$, we write for the right-hand side of (2.1.94) approximately

$$\begin{aligned} &\tfrac{1}{2}B^2(0)\left[B^2(\xi)-1\right]\cos 2\zeta\phi_{\Lambda'}(z) \\ &\quad +\tfrac{1}{2}B^2(0)\left[B^{-2}(\xi)-1\right]\left\{1-\tfrac{1}{2}\zeta^2(\xi\cdot\nabla\phi_\Lambda(z))^2\right\}. \end{aligned} \tag{2.1.95}$$

Here, we introduced the centre-of-mass coordinate z and the relative coordinate ξ

$$z = \tfrac{1}{2}(r + r'), \qquad \xi = r - r'. \tag{2.1.96}$$

In (2.1.87), up to second order in μ, we obtain

$$\begin{aligned} A'(\phi_{\Lambda'}) = \int \mathrm{d}^2 r \Big\{ &\frac{1}{2}\left(1 + \frac{\zeta^2\mu^2}{8}B^2(0)a_2\right)(\nabla\phi_{\Lambda'}(r))^2 \\ &-\mu B(0)\cos\zeta\phi_{\Lambda'}(r)\Big\}. \end{aligned} \tag{2.1.97}$$

Here, we defined

$$a_2 = \int \mathrm{d}^2\xi\cdot\xi^2\left[B^{-2}(\xi)-1\right]. \tag{2.1.98}$$

As will be mentioned later, the term $\cos 2\zeta\phi_{\Lambda'}$ can be ignored when Λ' is small. Therefore, it does not appear explicitely in (2.1.97).

Rescaling $\phi_{\Lambda'}$ in (2.1.97) like

$$\tilde{\phi}_{\Lambda'} = \phi_{\Lambda'}\sqrt{1 + \frac{\zeta^2\mu^2}{8}B^2(0)a_2}\,, \tag{2.1.99}$$

then we can write for (2.1.97)

$$S'(\tilde{\phi}_{\Lambda'}) = \int \mathrm{d}^2 r\left\{\frac{1}{2}(\nabla\tilde{\phi}_{\Lambda'}(r))^2 - \mu'\cos\zeta'\tilde{\phi}_{\Lambda'}(r)\right\}, \tag{2.1.100}$$

where we introduced the notation

$$\mu' = B(0)\mu\,, \tag{2.1.101a}$$

$$\zeta' = \zeta\Big/\sqrt{1 + \frac{\zeta^2\mu^2}{8}B^2(0)a_2}\,. \tag{2.1.101b}$$

We conclude that when the cutoff is shifted from Λ to Λ', the parameters appearing in the action are shifted and become μ' and ξ'.

We discuss this parameter shift in more detail. First, we assume $\Lambda' - \Lambda$ to be infinitesimal:

$$\Lambda' = \Lambda + \mathrm{d}\Lambda = \Lambda - |\mathrm{d}\Lambda| . \qquad (2.1.102)$$

Then, $G_h(r)$ in (2.1.91) becomes

$$G_h(r) = \frac{|\mathrm{d}\Lambda|}{\Lambda}\frac{1}{(2\pi)^2}\int_0^{2\pi}\mathrm{d}\theta\,\mathrm{e}^{\mathrm{i}\Lambda|r|\cos\theta} = \frac{|\mathrm{d}\Lambda|}{\Lambda}\frac{1}{2\pi}J_0(\Lambda|r|) . \qquad (2.1.103)$$

$J_0(x)$ contains oscillation modes and decreases only slowly for $x \gg 1$. This poor feature arises because an abrupt cutoff for the p-components has been introduced. This can be avoided by introducing a continuous cutoff in the p-space. Here, we will not introduce the explicit technical details, but we consider $J_0(\Lambda|r|)$ as being replaced, as a result of this improvement, by a function $\tilde{J}_0(\Lambda|r|)$ that decreases strongly for $|r| \gg \Lambda^{-1}$:

$$G_h(r) = \frac{|\mathrm{d}\Lambda|}{\Lambda}\frac{1}{2\pi}\tilde{J}_0(\Lambda|r|) . \qquad (2.1.104)$$

Using the same improvement, a_2 in (2.1.98) becomes

$$a_2 = \int \mathrm{d}^2\xi \cdot \xi^2 \cdot \zeta^2 G_h(\xi) = \alpha_2\zeta^2\frac{|\mathrm{d}\Lambda|}{\Lambda^5} , \qquad (2.1.105)$$

$$\alpha_2 = \int_0^{\infty}\mathrm{d}\rho \cdot \rho^3\tilde{J}_0(\rho) . \qquad (2.1.106)$$

Furthermore

$$B(0) = 1 - \frac{1}{2}\zeta^2 G_h(0) = 1 + \frac{1}{4\pi}\zeta^2\frac{\mathrm{d}\Lambda}{\Lambda} . \qquad (2.1.107)$$

Combining with (2.1.101a), we obtain

$$\mathrm{d}\mu = \mu' - \mu = (B(0) - 1)\mu = \frac{1}{4\pi}\zeta^2\mu\frac{\mathrm{d}\Lambda}{\Lambda} . \qquad (2.1.108)$$

From (2.1.101b), we obtain

$$\mathrm{d}\zeta = \zeta' - \zeta = -\frac{\zeta^2\mu^2}{16}B(0)^2\alpha_2\zeta = \frac{\alpha_2}{16}\zeta^5\mu^2\frac{\mathrm{d}\Lambda}{\Lambda^5} . \qquad (2.1.109)$$

Noting that owing to the definition of μ below (2.1.82), its dimension is given by $(\text{length})^2$, and we can define the dimensionless number y as

$$y = \mu\Lambda^{-2} . \qquad (2.1.110)$$

Defining x by

$$\zeta^2 = 8\pi\eta = 4\pi(2 + x) , \qquad (2.1.111)$$

equations (2.1.108) and (2.1.109) become

$$\frac{d \ln y}{d \ln \Lambda} = x \,, \tag{2.1.112a}$$

$$\frac{dx}{d \ln \Lambda} = 8\alpha_2 y^2 \,. \tag{2.1.112b}$$

Redefining $8\alpha_2 y^2$ as new variable y^2, finally we obtain the differential equation:

$$\frac{dy^2}{d \ln \Lambda} = 2xy^2 \,, \tag{2.1.113a}$$

$$\frac{dx}{d \ln \Lambda} = y^2 \,. \tag{2.1.113b}$$

The flow of this differential equation in the xy plane for the case that Λ is small is shown in Fig. 2.1. In particular, from (2.1.113),

$$\frac{d}{d \ln \Lambda}(x^2 - y^2) = 0 \tag{2.1.114}$$

can be deduced. Therefore, each trajectory is given by $x^2 - y^2 =$ constant. Setting $x = y$, (2.1.113) becomes

$$\frac{dx}{d \ln \Lambda} = x^2 \,, \tag{2.1.115}$$

and the solution can be given immediately:

$$x(\Lambda) = y(\Lambda) = \frac{x(\Lambda_0)}{1 + x(\Lambda_0) \ln \dfrac{\Lambda_0}{\Lambda}} \,. \tag{2.1.116}$$

Notice that $x(\Lambda) \simeq [\ln(\Lambda_0/\Lambda)]^{-1} \to 0$ for $\Lambda \to 0$.

Let us apply these considerations to the XXZ model. Following (2.1.111), x is given by $x = 2(\eta - 1)$, and we conclude therefore that long-range limit of the Heisenberg model discussed in the first half of this chapter just corresponds to the origin $x = y = 0$. Therefore the scaling flow for the Heisenberg model is on the line $x = y$, where the cut off dependence (2.1.116) should reappear. For the case of an anisotropy in the XY direction, η will be larger than $n = 1$ for the Heisenberg model, and y flows to $y(\Lambda) \to 0$. As expected, in the long-range limit the Umklapp-scattering can be neglected, no gap in the energy spectrum occurs, and the systems become critical.

On the other hand, in the case of an Ising-like anisotropy, η is smaller than one, and y flows to large values. The perturbative renormalization group equation described above becomes inadequate for the description. However, a large $y(\Lambda)$ can be interpreted as dominance of the $\cos 2\theta_+$ term, leading to a localized ground state for θ_+ at zero or at π. This corresponds to the

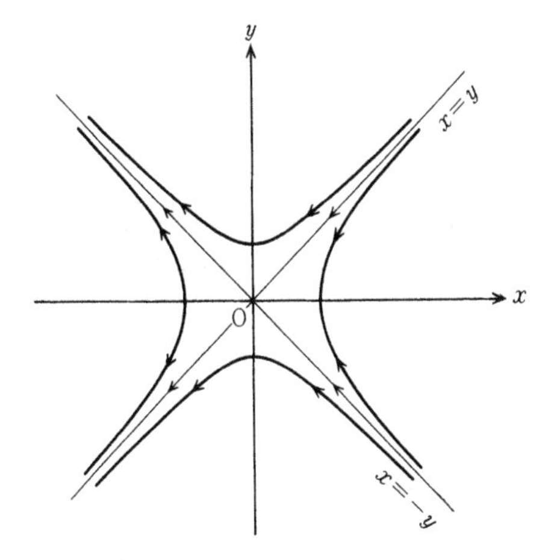

Fig. 2.1. The flow of the renormalization group in the Sine–Gordon model

charge density wave in the spinless fermion model, or in terms of the original spin model, the anti-ferromagnetic long-range order.

We will close this section on bosonization by providing an introduction to the problem of impurities. We consider the problem where at the origin $x = 0$ an impurity is present where the fermions scatter from the left to the right or vice versa. The Hamiltonian describing this process is given by

$$\begin{aligned} H_{\mathrm{imp}} &= \tilde{V}_0\big(R^\dagger(0)L(0) + L^\dagger(0)R(0)\big) \\ &= \int \mathrm{d}x\, \tilde{V}_0\delta(x)\big(R^\dagger(x)L(x) + L^\dagger(x)R(x)\big)\,. \end{aligned} \tag{2.1.117}$$

Performing the bosonization using (2.1.30), the Lagrangian L in the imaginary time formalism becomes

$$\begin{aligned} L &= L_0 + L_{\mathrm{imp}} \\ &= \int \mathrm{d}x \left[\frac{1}{4\pi\eta}\left(\frac{1}{v}(\partial_\tau\theta_+(x))^2 + v(\partial_x\theta_+(x))^2\right) + V_0\delta(x)\cos\theta_+(x)\right]. \end{aligned} \tag{2.1.118}$$

Here, we defined $V_0 = \tilde{V}_0/(\pi\alpha)$. Because the only non-linear term in this Lagrangian is expressed by $\theta_+(x=0)$, we can perform the integration with respect to the other coordinates $\theta_+(x \neq 0)$ and deduce an effective Lagrangian in terms of $\theta_+(x=0)$. That is, $\theta_+(x \neq 0)$ can be considered as being a heat bath. Therefore, we perform the integration under the constraint condition $\theta(\tau) = \theta_+(x=0,\tau)$. That is, we deduce $A_{\mathrm{eff}}(\theta(\tau))$ by

$$\begin{aligned} Z &= \int \mathcal{D}\theta_+(x,\tau)\, \mathrm{e}^{-\int_0^\beta \mathrm{d}\tau\, L} \\ &= \int \mathcal{D}\theta_+(x,\tau)\mathcal{D}\theta(\tau)\delta(\theta(\tau) - \theta_+(0,\tau))\, \mathrm{e}^{-\int_0^\beta \mathrm{d}\tau\, L} \end{aligned}$$

$$= \int \mathcal{D}\theta_+(x,\tau)\mathcal{D}\theta(\tau)\mathcal{D}\lambda(\tau)\,\mathrm{e}^{-\int_0^\beta \mathrm{d}\tau\,[L+\lambda(\tau)(\theta(\tau)-\theta_+(0,\tau))]}$$

$$= \int \mathcal{D}\theta(\tau)\,\mathrm{e}^{-A_{\mathrm{eff}}(\theta(\tau))}\,. \qquad (2.1.119)$$

Here, we expressed $\delta(\theta(\tau)-\theta_+(0,\tau))$ by introducing the Lagrange multiplier $\lambda(\tau)$. Then, the $\theta_+(x,\tau)$ integration can be performed without any constraint condition. Furthermore, in $L = L_0(\theta_+) + L_{\mathrm{imp}}(\theta_+(0,\tau))$, $\theta_+(0,\tau)$ can be replaced by $\theta(\tau)$; the only term containing $\theta_+(x,\tau)$ in the path integral is given by

$$\int_0^\beta \mathrm{d}\tau\,[L_0 - \lambda(\tau)\theta_+(0,\tau)]$$

$$= \sum_{\mathrm{i}\omega_n}\sum_q \left[\frac{\omega_n^2+v^2q^2}{4\pi\eta v}\theta_+(q,\mathrm{i}\omega_n)\theta_+(-q,-\mathrm{i}\omega_n)\right.$$

$$\left. - \frac{1}{2\sqrt{L}}(\lambda(\mathrm{i}\omega_n)\theta_+(-q,-\mathrm{i}\omega_n) + \theta_+(q,\mathrm{i}\omega_n)\lambda(-\mathrm{i}\omega_n))\right]. \qquad (2.1.120)$$

Because this is quadratic in θ_+, the integral is Gaussian and can be performed by completing the square. The result is an action in terms of λ given by

$$-\frac{1}{L}\sum_{\mathrm{i}\omega_n}\sum_q \frac{\pi\eta v}{\omega_n^2+v^2q^2}\lambda(\mathrm{i}\omega_n)\lambda(-\mathrm{i}\omega_n)\,. \qquad (2.1.121)$$

The sum in q becomes in the limit $L \to \infty$ an integral

$$\frac{1}{L}\sum_q \frac{\pi\eta v}{\omega_\eta^2+v^2q^2} = \int \frac{\mathrm{d}q}{2\pi}\frac{\pi\eta v}{\omega_n^2+v^2q^2} = \frac{\pi\eta}{2|\omega_n|}\,. \qquad (2.1.122)$$

Therefore, the terms containing λ in the action are given by

$$\sum_{\mathrm{i}\omega_n}\left[-\frac{\pi\eta}{2|\omega_n|}\lambda(\mathrm{i}\omega_n)\lambda(-\mathrm{i}\omega_n) + \frac{1}{2}[\lambda(\mathrm{i}\omega_n)\theta(-\mathrm{i}\omega_n) + \theta(\mathrm{i}\omega_n)\lambda(-\mathrm{i}\omega_n)])\right]. \qquad (2.1.123)$$

Again, by completing the square we obtain

$$\sum_{\mathrm{i}\omega_n}\frac{|\omega_n|}{2\pi\eta}\theta(\mathrm{i}\omega_n)\theta(-\mathrm{i}\omega_n)\,. \qquad (2.1.124)$$

This term describes the effect of the heat bath on $\theta(\tau) = \theta_+(0,\tau)$. The appearance of $|\omega_n|$ signifies the dissipation or friction, describing an irreversible process. Re-expressing ω_n in terms of τ-integrals, we obtain a double integral in τ, τ', signifying that a non-local interaction takes place along the time direction. Finally, the effective action A_{eff} with respect to $\theta(\tau)$ becomes

$$A_{\text{eff}} = \sum_{\mathrm{i}\omega_n} \frac{|\omega_n|}{2\pi\eta} \theta(\mathrm{i}\omega_n)\theta(-\mathrm{i}\omega_n) + V_0 \int_0^\beta \mathrm{d}\tau \cos\theta(\tau) . \qquad (2.1.125)$$

We reduced the problem in this manner to the single-particle problem moving in the potential $V_0 \cos\theta$ subject to dissipation with friction constant $\gamma = 1/(2\pi\eta)$. As described in Sect. 5.2 of ref. [G.1], this single-particle motion corresponds to a Bloch wave (band motion) in the case when the friction is small. When the friction is larger than the critical value $\gamma_c = 1/(4\pi)$, translational symmetry breaks down, and the particle comes to rest in some minimum of the potential. In terms of the original fermion problem, we obtain the following picture. As can be seen from (2.1.25a), when $\theta(\tau) = \theta_+$ is changed about 2π, this corresponds to the transition of one fermion from the left to the right side. Therefore, the Bloch state of $\theta(\tau)$ corresponds to the perfect transmission of the fermion through the potential caused by the impurity. On the other hand, the localized state corresponds to a total reflection. The former corresponds to the case $\eta > 2$ of an attractive interaction; the latter to the case $\eta < 2$ of a repulsive interaction. The case when $\eta = 2$ (where the fermions do not interact) is special. Then, $V_0 \cos\theta$ becomes marginal. Depending on V_0, both transmission and reflection occur at the impurity. The above description can be understood from the fact that η is the parameter that determines the competition between the particle-like and wave-like nature of the system. In such a manner, in the one-dimensional Tomonaga–Luttinger liquid, being a non-Fermi liquid, the effect of impurities leads to a singular behaviour that cannot be seen in the Fermi liquid.

2.2 Conformal Field Theory

In this section, we will discuss the conformal field theory, which is a powerful tool to study the critical properties of one-dimensional quantum systems. Conformal field theory is a field theory that is invariant under conformal mapping (angle conserving projections) in the theory of complex functions.

Owing to this high symmetry, many features of the properties of the spectrum of such a system are already determined. As a result, for example, the asymptotic behaviour of the correlation functions is determined just by the energy eigenvalues of the finite system. Therefore, this method offers powerful tools to study the one-dimensional quantum systems in combination with numerical calculations or the Bethe ansatz. Here, we will not give a systematic introduction, but we will introduce this approach by applying it to the action (2.1.73)

$$A_0 = \int \mathrm{d}\tau \int \mathrm{d}x \, \frac{1}{4\pi\eta} \left[\frac{1}{v} (\partial_\tau \theta_+)^2 + v(\partial_x \theta_+)^2 \right] . \qquad (2.2.1)$$

For simplicity, we introduce the scaling factors

$$\theta_+ = \sqrt{2\eta}\,\phi\,, \tag{2.2.2}$$

$$r_0 = v\tau\,, \tag{2.2.3}$$

$$r_1 = r\,. \tag{2.2.4}$$

In these variables, A_0 can be expressed as

$$A_0 = \int_S \frac{\mathrm{d}^2 r}{2\pi} (\nabla\phi)^2\,. \tag{2.2.5}$$

Here, the region of integration S in the (r_0, r_1) plane is given by $[0, v\beta] \times [0, L]$ (L is the length of the system). In what follows, we will for a while consider a general region S. Recall that in (2.2.1) and (2.2.5), right-moving and left-moving components are present. That is, considering the equations of motion derived from (2.2.1) in the real space

$$\left(v^2\partial_t^2 - \partial_x^2\right)\theta_+(x,t) = 0\,, \tag{2.2.6}$$

the solution can in general be expressed by a right- and a left-moving part ϕ_R and ϕ_L:

$$\theta_+(x,t) = \phi_\mathrm{R}(x - vt) + \phi_\mathrm{L}(x + vt)\,. \tag{2.2.7}$$

Interpreting these equations as the time development of (2.1.24a), ϕ_R and ϕ_L become the phase field of the right-moving and left-moving fermions, respectively. Because we want to consider the imaginary time formalism, we rewrite (2.2.7) as

$$\phi(r) = \bar{\varphi}(r_0 - \mathrm{i}r_1) + \varphi(r_0 + \mathrm{i}r_1)\,. \tag{2.2.8}$$

Introducing complex coordinates $\xi = r_0 + \mathrm{i}r_1$ and $\bar{\xi} = r_0 - ir_0$, $\varphi(\xi)$ is a function only of ξ, and $\bar{\varphi}(\bar{\xi})$ is a function only of $\bar{\xi}$. The property 'being only a function of ξ' is well known in complex analysis and just means 'analytic function'. Therefore, many beautiful results concerning analytic functions can be applied to $\varphi(\xi)$ as well as to $\bar{\varphi}(\bar{\xi})$.

In a word, conformal field theory is a field theory employing analytic functions. Here, the conformal mapping from one complex plane to another using holomorphic functions is of special interest. The most important conformal transformation is given by

$$z(\xi) = \mathrm{e}^{2\pi\xi/L}\,. \tag{2.2.9}$$

Considering the limit $\beta \to \infty$ of zero temperature, the region $[-\beta/2, \beta/2]$ becomes $[-\infty, +\infty]$, and in the spatial region we consider a system with size L and impose periodic boundary conditions ($x \in [0, L]$). Then, ξ defined in the region $[-\infty, +\infty] \times [0, L]$ is projected to the whole z-plane as given by (2.2.9). The radial direction on the z plane corresponds to time development, and the angle corresponds to the spatial coordinate.

The following discussion holds also in general. We start by discussing the properties of operators under conformal transformations. Introducing the transformation of the coordinates

$$r_\mu \to r'_\mu = r_\mu + \delta r_\mu \,,$$
$$\frac{\partial}{\partial r_\mu} \to \frac{\partial}{\partial r'_\mu} = \left(\delta_{\mu\nu} - \frac{\partial \delta r_\nu}{\partial r_\mu}\right)\frac{\partial}{\partial r_\nu}\,, \tag{2.2.10}$$

the energy–momentum tensor is defined by the change of the action under this transformation

$$A \to A + \delta A = A - \frac{1}{2\pi}\int \mathrm{d}^2 r \sum_{\mu,\nu} T_{\mu\nu}(r)\partial_\mu(\delta r_\nu) \tag{2.2.11}$$

[see App. B]. Considering explicitly the action A_0 (2.2.5), we obtain

$$T_{\mu\nu}(r) = \,:2\partial_\mu\phi\partial_\nu\phi - \delta_{\mu\nu}\sum_\lambda(\partial_\lambda\phi)^2: \tag{2.2.12}$$

The requirement that the action is invariant under the transformation δr_μ restricts the possible form of $T_{\mu\nu}(r)$.

First, considering linear transformations $\delta r_\mu = a_\mu = \text{const}$. From $\delta A = 0$, we obtain by partial integration

$$\partial_\mu T_{\mu\nu}(r) = 0\,. \tag{2.2.13}$$

In the same manner, from rotational symmetry, we obtain with $\delta r_\mu = \varepsilon_{\mu\nu} r_\nu$,

$$T_{\mu\nu}(r) = T_{\nu\mu}(r)\,, \tag{2.2.14}$$

and from scaling invariance $\delta r_\mu = \varepsilon r_\mu$,

$$\sum_\mu T_{\mu\mu}(r) = 0\,. \tag{2.2.15}$$

It is simple to check that all these equations are satisfied for (2.2.12) (Notice that the equation of motion $\nabla^2\phi(r) = 0$ must be used).

We can split $T_{\mu\nu}$ into holomorphic and anti-holomorphic parts $T(z)$ and $\bar{T}(\bar{z})$. Owing to (2.2.14) and (2.2.15), $T_{\mu\nu}$ can be expressed by two independent components T_{00} and T_{01}. Defining

$$T = -\tfrac{1}{2}(T_{00} - \mathrm{i}T_{01})\,, \qquad \bar{T} = -\tfrac{1}{2}(T_{00} + \mathrm{i}T_{01})\,, \tag{2.2.16}$$

from (2.2.13) we deduce that

$$\partial_{\bar{z}} T = 0\,, \qquad \partial_z \bar{T} = 0\,. \tag{2.2.17}$$

That is, the z-dependence is given by $T = T(z)$, $\bar{T} = \bar{T}(\bar{z})$.

In the same manner as in (2.2.12), we obtain

$$\begin{aligned} T(z) &= -\tfrac{1}{2}\left[:(\partial_0\phi)^2 - (\partial_1\phi)^2 - 2\mathrm{i}(\partial_0\phi)(\partial_1\phi):\right] \\ &= -2:(\partial_z\phi)^2:, \end{aligned} \tag{2.2.18}$$

and

$$\bar{T}(\bar{z}) = -2:(\partial_{\bar{z}}\phi)^2:.$$

As demonstrated, when the transformation properties of the action are determined, the transformation properties of the operators can be deduced. Now, we consider the correlation function of the operators $\mathcal{O}_1, \mathcal{O}_2, \ldots, \mathcal{O}_N$:

$$\langle \mathcal{O}_1(r_1)\ldots\mathcal{O}_N(r_N)\rangle = \int \mathcal{D}\phi\, \mathrm{e}^{-A}\mathcal{O}_1(r_1)\ldots\mathcal{O}_N(r_N). \tag{2.2.19}$$

Under the coordinate transformation (2.2.10), the operator changes $\mathcal{O}_i(r) \to \mathcal{O}_i(r) + \delta\mathcal{O}_i(r)$. Requiring that the correlation function is invariant under this transformation, we obtain

$$\begin{aligned} &\langle \mathcal{O}_1(r_1)\ldots\mathcal{O}_N(r_N)\rangle \\ &\quad= \int \mathcal{D}\phi\, \mathrm{e}^{-A-\delta A}(\mathcal{O}_1(r_1)+\delta\mathcal{O}_1(r_1))\ldots(\mathcal{O}_N(r_N)+\delta\mathcal{O}_N(r_N)) \\ &\quad= \langle \mathcal{O}_1(r_1)\ldots\mathcal{O}_N(r_N)\rangle - \int \mathcal{D}\phi\, \mathrm{e}^{-A}\delta A\mathcal{O}_1(r_1)\ldots\mathcal{O}_N(r_N) \\ &\qquad+ \int \mathcal{D}\phi\, \mathrm{e}^{-A}\sum_{i=1}^{N}\mathcal{O}_1(r_1)\ldots\delta\mathcal{O}_i(r_i)\ldots\mathcal{O}_N(r_N) \end{aligned} \tag{2.2.20}$$

and finally the equation

$$\begin{aligned} &\sum_{i=1}^{N}\langle \mathcal{O}_1(r_1)\ldots\delta\mathcal{O}_i(r_i)\ldots\mathcal{O}_N(r_N)\rangle \\ &\qquad+ \int \frac{\mathrm{d}^2y}{2\pi}(\partial_\mu\delta r_\nu(y))\langle T_{\mu\nu}(y)\mathcal{O}_1(r_1)\ldots\mathcal{O}_N(r_N)\rangle = 0. \end{aligned} \tag{2.2.21}$$

Introducing the notation $X \equiv \mathcal{O}_1(r_1)\ldots\mathcal{O}_N(r_N)$, the second term on the left-hand side of (2.2.21) becomes

$$\begin{aligned} &\int \frac{\mathrm{d}^2y}{2\pi}(\partial_\mu\delta r_\nu(y))\langle T_{\mu\nu}(y)X\rangle \\ &\qquad= \int \frac{\mathrm{d}^2y}{2\pi}\{\partial_\mu[\delta r_\nu(y)\langle T_{\mu\nu}(y)X\rangle] - \delta r_\nu(y)\langle(\partial_\mu T_{\mu\nu}(y))X\rangle\}. \end{aligned} \tag{2.2.22}$$

The second term vanishes owing to (2.2.11). Introducing $z = y_0 + \mathrm{i}y_1$ and defining $\varepsilon(z) = \delta y_0 + \mathrm{i}\delta y_1$ and $\bar{\varepsilon}(\bar{z}) = \delta y_0 - \mathrm{i}\delta y_1$, the first term can be expressed using $T(z)$ and $\bar{T}(\bar{z})$ (2.2.16) as

$$\int \frac{\mathrm{d}^2y}{2\pi}\{\partial_{\bar{z}}[\varepsilon(z)\langle T(z)X\rangle] + \partial_z[\bar{\varepsilon}(\bar{z})\langle\bar{T}(\bar{z})X\rangle]\}. \tag{2.2.23}$$

Here the reader might think that because $\varepsilon(z)\langle T(z)X\rangle$ is a function of z only, the term vanishes when $\partial_{\bar{z}}$ is applied. However, here it is necessary to recall the following equation

$$\frac{\partial}{\partial \bar{z}}\frac{1}{z} = \frac{\partial}{\partial z}\frac{1}{\bar{z}} = \pi\delta^2(r)\,. \tag{2.2.24}$$

This equation can be deduced by considering

$$\frac{1}{z} = \lim_{a\to 0}\frac{\bar{z}}{z\bar{z}+a^2}\,, \tag{2.2.25}$$

and taking the derivative with respect to $\bar{z}$. We obtain

$$\frac{\partial}{\partial \bar{z}}\frac{\partial}{\partial z}\ln z = \frac{\partial}{\partial \bar{z}}\frac{\partial}{\partial z}\ln \bar{z} = \pi\delta^2(r)\,. \tag{2.2.26}$$

Defining the Green function of the Laplacian operator with suitable boundary conditions

$$-\nabla^2 G(r,r') = -\frac{\partial}{\partial z}\frac{\partial}{\partial \bar{z}}G(r,r') = \delta^2(r-r') \tag{2.2.27}$$

in the infinite plane (being the z-plane introduced by (2.2.9)) we obtain

$$G(r,r') = \frac{1}{\pi}\langle\phi(r)\phi(r')\rangle = -\frac{1}{4\pi}\ln \bar{z}z = -\frac{1}{2\pi}\ln|z|\,. \tag{2.2.28}$$

In much the same way, a singularity in $\langle T(z)X\rangle$ may arise when acting on it with $\partial_{\bar{z}}$. That is, writing $w_1,\ldots,w_N$ for the coordinates $r_1,\ldots,r_N$, with m being an integer, as will be described in what follows, singularities of the form

$$\frac{1}{(z-w_i)^m} \tag{2.2.29}$$

will arise in $\langle T(z)X\rangle$. Then, in the Laurent expansion

$$\varepsilon(z)\frac{1}{(z-w)^m} = \sum_{l=0}^{\infty}\frac{1}{l!}\frac{\partial^l\varepsilon(w)}{\partial w^l}(z-w)^{l-m} \tag{2.2.30}$$

when acting with $\partial_{\bar{z}}$, owing to (2.2.24), only the term with $l-m=-1$ contributes to the integration, and all the other terms vanish. This can be understood by considering, for example,

$$\frac{\partial}{\partial \bar{z}}\frac{1}{z^2} = \frac{2}{z}\frac{\partial}{\partial \bar{z}}\frac{1}{z} = \lim_{a\to 0}\frac{2(x-iy)}{x^2+y^2+a^2}\cdot 2\pi\delta^2(r) = 0\,. \tag{2.2.31}$$

This just corresponds to the residue theorem. Therefore, instead of the two-dimensional integral in (2.2.23), we can rewrite the expression using a complex integral around the contour C surrounding all points $w_1,\ldots,w_N$

$$\oint_C \frac{dz}{2\pi i}\varepsilon(z)\langle T(z)X\rangle\,. \tag{2.2.32}$$

Now we discuss how singularities of the form (2.2.29) might arise in $\langle T(z)X\rangle$. To do so, we must consider the operator product expansion (OPE). We start with Wick's theorem. In the z-plane introduced by (2.2.9), where the radial direction corresponds to the time direction, we define the radial ordering operator R as follows

$$RA(z)B(w) = \begin{cases} A(z)B(w) & (|z| > |w|) \\ B(w)A(z) & (|z| < |w|) \end{cases}\,. \tag{2.2.33}$$

Then, Wick's theorem gives for the product of two operators

$$RA(z)B(w) = \overbrace{A(z)B}(w) + :A(z)B(w):\,. \tag{2.2.34}$$

Here, we defined the so-called contraction by

$$\overbrace{A(z)B}(w) = \langle 0|RA(z)B(w)|0\rangle\,. \tag{2.2.35}$$

As an example, we consider

$$\begin{aligned} A(z) &= \partial_z\phi(z) \\ B(w) &= \partial_w\phi(w)\,. \end{aligned} \tag{2.2.36}$$

We obtain

$$\begin{aligned} \partial_z\overbrace{\phi(z)\partial_w\phi}(w) &= \partial_z\partial_w\overbrace{\phi(z)\phi}(w) \\ &= -\partial_z\partial_w\left[\frac{1}{2}\ln|z-w|\right] = -\frac{1}{4}\frac{1}{(z-w)^2}\,. \end{aligned} \tag{2.2.37}$$

This equation is exact. Using this, we obtain, for example,

$$\begin{aligned} RT(z)\partial_w\phi(w) &= R\left[-2:(\partial_z\phi(z))^2:\partial_w\phi(w)\right] \\ &= -4\partial_z\overbrace{\phi(z)\partial_w\phi}(w):\partial_z\phi(z): - 2:(\partial_z\phi(z))^2\partial_w\phi(w): \\ &= \frac{:\partial_z\phi(z):}{(z-w)^2} - 2:(\partial_z\phi(z))^2\partial_w\phi(w):\,. \end{aligned} \tag{2.2.38}$$

For the case when $\partial_w\phi(z)$ is contained in $\mathcal{O}_i$, the singularity $(z-w)^2$ arises.

We obtain for $z \to z + \varepsilon(z)$ the transformation law

$$\delta_\varepsilon\mathcal{O}(z) = \oint_C \frac{d\xi}{2\pi i}\varepsilon(\zeta)T(\zeta)\mathcal{O}(z) \tag{2.2.39}$$

of the operator $\mathcal{O}(z)$. Here, C is a contour that encircles z. From (2.2.38), we obtain for $\partial_z\phi(z)$ (omitting : :)

$$\delta_\varepsilon(\partial_z\phi(z)) \equiv T_\varepsilon(\partial_z\phi(z)) = \oint \frac{d\zeta}{2\pi i}\varepsilon(\zeta)\left[\frac{\partial_\zeta\phi(\zeta)}{(\zeta-z)^2} + (\text{regular})\right]\,. \tag{2.2.40}$$

Due to the Cauchy-theorem, the regular part does not contribute the integral. The singular part becomes

$$\frac{\partial_\zeta \phi(\zeta)}{(\zeta - z)^2} = \frac{\partial_z \phi(z)}{(\zeta - z)^2} + \frac{\partial_z^2 \phi(z)}{(\zeta - z)} + (\text{regular}) \tag{2.2.41}$$

and therefore, (2.2.40) becomes

$$\delta_\varepsilon(\partial_z \phi(z)) = (\partial_z \varepsilon(z))\partial_z \phi(z) + \varepsilon(z)\partial_z(\partial_z \phi(z)) \,. \tag{2.2.42}$$

This is the transformation rule for infinitesimal transformations, and owing to the integral we conclude for a general conformal transformation $z \to w = w(z)$

$$\partial_z \phi(z) \to \partial_w \phi(w) \frac{\mathrm{d}w}{\mathrm{d}z} \,. \tag{2.2.43}$$

Indeed, when setting $w = z + \varepsilon(z)$ ($\varepsilon(z)$ infinitesimal), we immediately regain (2.2.42). In general, an operator $\mathcal{O}(z, \bar{z})$ transforming under $z \to w = w(z)$ and $\bar{z} \to \bar{w} = \bar{w}(\bar{z})$ as

$$\mathcal{O}(z, \bar{z}) \to \tilde{\mathcal{O}}(w, \bar{w}) = \mathcal{O}(w, \bar{w}) \left(\frac{\mathrm{d}w}{\mathrm{d}z}\right)^{\Delta} \left(\frac{\mathrm{d}\bar{w}}{\mathrm{d}\bar{z}}\right)^{\bar{\Delta}} \tag{2.2.44}$$

is called a primary field. $(\Delta, \bar{\Delta})$ are the so-called conformal dimensions. For such a field,

$$R[T(\zeta)\mathcal{O}(z, \bar{z})] = \frac{\Delta}{(\zeta - z)^2}\mathcal{O}(z, \bar{z}) + \frac{1}{\zeta - z}\partial_z \mathcal{O}(z, \bar{z}) + (\text{regular}) \tag{2.2.45}$$

and

$$\delta \mathcal{O}(z, \bar{z}) = \Delta(\partial_z \varepsilon(z))\mathcal{O}(z, \bar{z}) + \varepsilon(z)\partial_z \mathcal{O}(z, \bar{z}) \tag{2.2.46}$$

hold. For the anti-holomorphic part, similar equations hold.

In such a way, the operator product expansion with the energy-momentum tensor $T(z)$ and $\bar{T}(\bar{z})$ determines the transformation properties of the operator. So, what might be the transformation properties of $T(z)$ and $\bar{T}(\bar{z})$ themselves? What we find is that $T(z)$ and $\bar{T}(\bar{z})$ are not primary fields. After some calculations, for (2.2.18) we obtain

$$RT(z)T(w) = \frac{c/2}{(z - w)^4} + \frac{2T(w)}{(z - w)^2} + \frac{\partial_w T(w)}{z - w} + (\text{regular}) \,. \tag{2.2.47}$$

Here, c is the so-called central charge, being in the present case $c = 1$. This number can in some sense be interpreted as the number of bosonic components. The transformation law corresponding to this OPE is given by

$$\delta_\varepsilon T(z) = \varepsilon(z)(\partial_z T(z)) + 2(\partial_z \varepsilon(z))T(z) + \frac{c}{12}\partial_z^3 \varepsilon(z) \,. \tag{2.2.48}$$

For general $z \to w(z)$, the transformation law is

$$T(z) \to \tilde{T}(w) = T(w)\left(\frac{\mathrm{d}w}{\mathrm{d}z}\right)^2 + \frac{c}{12}\{w, z\}\,. \tag{2.2.49}$$

Here, $\{w, z\}$ is the so-called Schwartz derivative, defined as

$$\{w, z\} = \frac{\mathrm{d}^3 w}{\mathrm{d}z^3}\Big/\frac{\mathrm{d}w}{\mathrm{d}z} - \frac{3}{2}\left(\frac{\mathrm{d}^2 w}{\mathrm{d}z^2}\right)^2\Big/\left(\frac{\mathrm{d}w}{\mathrm{d}z}\right)^2. \tag{2.2.50}$$

Now, we introduce the following expansion for $T(z)$ and $\bar{T}(\bar{z})$

$$\begin{aligned} T(z) &= \sum_{n=-\infty}^{\infty} z^{-n-2} L_n\,, \\ \bar{T}(\bar{z}) &= \sum_{n=-\infty}^{\infty} \bar{z}^{-n-2} \bar{L}_n\,. \end{aligned} \tag{2.2.51}$$

Conversely, we obtain

$$\begin{aligned} L_n &= \oint_C \frac{\mathrm{d}z}{2\pi \mathrm{i}} z^{n+1} T(z)\,, \\ \bar{L}_n &= \oint_C \frac{\mathrm{d}\bar{z}}{2\pi \mathrm{i}} \bar{z}^{n+1} \bar{T}(\bar{z})\,. \end{aligned} \tag{2.2.52}$$

Here, C is an arbitrary loop encircling the origin. We derive the commutation relation that L_n obeys. C_w is defined as a path surrounding the point w

$$\begin{aligned} [L_m, L_n] &= \oint_{c_1} \frac{\mathrm{d}z}{2\pi \mathrm{i}} \oint_{c_2} \frac{\mathrm{d}w}{2\pi \mathrm{i}} z^{m+1} w^{h+1} (T(z)T(w) - T(w)T(z)) \\ &= \oint \frac{\mathrm{d}z}{2\pi \mathrm{i}} \oint_{|z|>|w|} \frac{\mathrm{d}w}{2\pi \mathrm{i}} z^{m+1} w^{n+1} R[T(z)T(w)] \\ &\quad - \oint \frac{\mathrm{d}z}{2\pi \mathrm{i}} \oint_{|z|<|w|} \frac{\mathrm{d}w}{2\pi \mathrm{i}} z^{m+1} w^{n+1} R[T(z)T(w)] \\ &= \oint_c \frac{\mathrm{d}w}{2\pi \mathrm{i}} \left(\oint_{|z|>|w|} \frac{\mathrm{d}z}{2\pi \mathrm{i}} - \oint_{|z|<|w|} \frac{\mathrm{d}z}{2\pi \mathrm{i}} \right) z^{m+1} w^{n-1} R[T(z)T(w)] \\ &= \oint_c \frac{\mathrm{d}w}{2\pi \mathrm{i}} \oint_{c_w} \frac{\mathrm{d}z}{2\pi \mathrm{i}} z^{m+1} w^{n+1} \\ &\quad \times \left[\frac{c/2}{(z-w)^4} + \frac{2T(w)}{(z-w)^2} + \frac{\partial_w T(w)}{z-w} + (\text{regular}) \right] \\ &= \oint_c \frac{\mathrm{d}w}{2\pi \mathrm{i}} w^{n+1} \left[\frac{c}{12}(m+1)m(m-1) w^{m-2} \right. \\ &\quad \left. + 2(m+1) w^m T(w) + w^{m+1} \partial_w T(w) \right] \end{aligned}$$

$$= \frac{c}{12}(m^3 - m)\delta_{n+m,0} + 2(m+1)L_{n+m} - (n+m+2)L_{n+m}$$
$$= \frac{c}{12}(m^3 - m)\delta_{n+m,0} + (m-n)L_{n+m}\,.$$

The result is

$$[L_m, L_n] = (m-n)L_{n+m} + \frac{c}{12}(m^3 - m)\delta_{n+m,0}\,. \tag{2.2.53}$$

The algebra obtained from these commutation relations is called Virasoro-algebra.

For $\mathcal{O}(z)$ being a primary field, we obtain from (2.2.45)

$$\begin{aligned} R[T(z)\mathcal{O}(0)] &= \frac{\Delta}{z^2}\mathcal{O}(0) + \frac{1}{z}\partial_z\mathcal{O}(0) + (\text{regular}) \\ &= \sum_{n=-\infty}^{\infty} \frac{1}{z^{n+2}} L_n\mathcal{O}(0)\,, \end{aligned} \tag{2.2.54}$$

and therefore

$$\begin{aligned} L_0\mathcal{O}(0) &= \Delta\mathcal{O}(0)\,, \\ L_n\mathcal{O}(0) &= 0 \qquad (n > 0)\,. \end{aligned} \tag{2.2.55}$$

That is, when L_0 acts on $\mathcal{O}(0)$ the conformal dimension Δ is obtained; when L_n $(n > 0)$ acts on $\mathcal{O}(0)$, the result vanishes. Setting $m = 0$ in (2.2.53), we obtain

$$[L_0, L_n] = -nL_n\,, \tag{2.2.56}$$

and therefore L_n $(n < 0)$ is an operator that raises the eigenvalue of L_0.

Defining the vacuum as

$$\mathrm{L}_n|\text{vac}\rangle \qquad (n \geqq -1)\,, \tag{2.2.57}$$

for the state

$$|\mathcal{O}\rangle = \mathcal{O}(0)|\text{vac}\rangle \tag{2.2.58}$$

the equation

$$L_0|\mathcal{O}\rangle = L_0\mathcal{O}(0)|\text{vac}\rangle = \Delta\mathcal{O}(0)|\text{vac}\rangle = \Delta|\mathcal{O}\rangle \tag{2.2.59}$$

holds, and we obtain

$$\begin{aligned} L_0(L_{-n}|\mathcal{O}\rangle) &= ([L_0, L_{-n}] + L_{-n}L_0)|\mathcal{O}\rangle \\ &= nL_{-n}|\mathcal{O}\rangle + \Delta L_{-n}|\mathcal{O}\rangle = (n+\Delta)L_{-n}|\mathcal{O}\rangle\,. \end{aligned} \tag{2.2.60}$$

That is, the eigenvalue of L_0 is raised about n when we act with the operator L_{-n}. This just corresponds to the ladder operator J^+ raising the eigenvalue of the momentum operator J^z.

Recalling that in the theory of angular momentum new states can be created just by using the commutation relations, we might expect that by acting with L_{-n}, a group of states can be built up. Indeed, starting from $|\mathcal{O}\rangle$, the states

$$L_{-n_1}L_{-n_2}\dots L_{-n_k}|\mathcal{O}\rangle \tag{2.2.61}$$

are eigenstates of L_0 with eigenvalue

$$\Delta+\sum_{j=1}^{k} n_j\,. \tag{2.2.62}$$

An analogous formula holds for the anti-holomorphic part.

Such a group of states can be built up for every primary field $\mathcal{O}$, and is called a conformal tower. This classification of states owing to conformal invariance is one interesting result of conformal field theory. Saying it the other way round, when the energy spectrum of a conformal invariant system is known, the conformal dimensions Δ and $\bar{\Delta}$ can be read off (there are, in general, a large number of conformal dimensions, if not infinitely many). When Δ is known, the asymptotic behaviour of the correlation function of the corresponding primary fields can be deduced easily. Now, we will apply these facts to the action (2.2.1) or (2.2.5) again.

First, notice that η does not appear in (2.2.5). η gives a relation between θ_+ and ϕ by the scaling (2.2.2), and this also induces the boundary conditions in (2.2.3). That is, the condition that θ_+ is 2π-periodic means that ϕ is $2\pi/\sqrt{2\eta}$-periodic. For a while, we will return to the real space and write this condition as

$$\phi(t,x+L)=\phi(t,x)+2\pi RN\,. \tag{2.2.63}$$

Here, $R=1/\sqrt{2\eta}$ is the compactification radius, and N is called the winding number. Keeping this in mind, writing $\phi(t,x)$ as

$$\phi(t,x)=\frac{2\pi RN}{L}x+\hat{\phi}(t,x)\,, \tag{2.2.64}$$

then $\hat{\phi}(t,x)$ is a periodic function.

Now, we decompose $\hat{\phi}(t,x)$ into its Fourier components.

$$\hat{\phi}(t,x)=\frac{1}{\sqrt{L}}\sum_{n=-\infty}^{\infty}\mathrm{e}^{\mathrm{i}k_n x}\hat{\phi}(t,k_n)\,, \tag{2.2.65}$$

with $k_n=2\pi n/L$. From the condition that $\hat{\phi}(t,x)$ is Hermitian, we obtain

$$\hat{\phi}^\dagger(t,k_n)=\hat{\phi}(t,-k_n)\,. \tag{2.2.66}$$

Next, we consider the canonical conjugated momentum $\Pi(t,x)$ of $\phi(x,t)$. From the Lagrangian, we deduce for Π that

$$\Pi(t,x) = \frac{\delta L}{\delta(\partial_t \phi(t,x))} = \frac{1}{\pi v}\partial_t \phi(t,x) . \tag{2.2.67}$$

Again, by expanding in Fourier components, we obtain

$$\Pi(t,x) = \frac{1}{\sqrt{L}} \sum_{n=-\infty}^{\infty} \mathrm{e}^{\mathrm{i}k_n x} \Pi(t,k_n) . \tag{2.2.68}$$

The equal-time commutator becomes

$$[\phi(t,x), \Pi(t,x')] = \frac{1}{L}\sum_{n,m} \mathrm{e}^{\mathrm{i}k_n x + \mathrm{i}k_m x'} \left[\hat{\phi}(t,k_n), \Pi(t,k_m)\right] . \tag{2.2.69}$$

By requiring that this becomes $\mathrm{i}\delta(x-x')$, we deduce that

$$\left[\hat{\phi}(t,k_n), \Pi(t,k_m)\right] = \mathrm{i}\delta_{k_n,-k_m} . \tag{2.2.70}$$

With γ_n being constants and $b_n^\dagger$, b_n being bosonic creation and annihilation operators, the commutation relation suggests writing

$$\begin{aligned} \hat{\phi}(t,k_n) &= \gamma_n \left(b_n + b_n^\dagger\right) , \\ \Pi(t,k_n) &= \frac{\mathrm{i}}{2\gamma_{-n}}\left(- b_n + b_{-n}^\dagger\right) . \end{aligned} \tag{2.2.71}$$

Notice that $\hat{\phi}^\dagger(t,k_n) = \hat{\phi}^\dagger(t,-k_n)$ and $\hat{\Pi}^\dagger(t,k_n) = \hat{\Pi}^\dagger(t,-k_n)$ are satisfied automatically, and that (2.2.70) holds.

Next, we insert this into the Hamiltonian

$$\begin{aligned} H &= \int \mathrm{d}x\, \Pi(x)\dot{\phi}(x) - L \\ &= \int \mathrm{d}x\, \pi v \Pi(x)^2 - \int \frac{\mathrm{d}x}{2\pi}\left(\pi^2 v \Pi(x)^2 - v(\nabla\phi(x))^2\right) \\ &= \int \mathrm{d}x \left[\frac{\pi}{2} v \Pi(x)^2 + \frac{v}{2\pi}(\nabla\phi(x))^2\right] , \end{aligned} \tag{2.2.72}$$

and obtain

$$H = \sum_{n=-\infty}^{\infty} \left(\frac{\pi}{2} v \Pi(k_n)\Pi(-k_n) + \frac{v k_n^2}{2\pi}\hat{\phi}(k_n)\hat{\phi}(-k_n)\right) + \frac{2\pi v}{L} R^2 N^2 , \tag{2.2.73}$$

and from the condition that the terms $b_n b_{-n}$ and $b_n^\dagger b_{-n}^\dagger$ cancel, we obtain

$$(\gamma_n \gamma_{-n})^2 = \left(\frac{\pi}{2k_n}\right)^2 . \tag{2.2.74}$$

For $n \neq 0$, we choose γ_n to be

$$\gamma_n = \gamma_{-n} = \sqrt{\frac{\pi}{2|k_n|}} . \tag{2.2.75}$$

On the other hand, $n = 0$ corresponds to the so-called zero mode. Setting $\hat{\phi}(k_n = 0) = \sqrt{L}Q$ and $\Pi(k_n = 0) = (1/\sqrt{L})P$, the Hamiltonian becomes

$$H = \frac{\pi v}{2L} P^2 + \sum_{n \neq 0} v|k_n| \left(b_n^\dagger b_n + \frac{1}{2} \right) + \frac{2\pi v}{L} R^2 N^2 , \tag{2.2.76}$$

P is a conserved quantity. Using the commutator $[Q, P] = \mathrm{i}$, we obtain

$$Q(t) = Q + \frac{\pi P}{L} vt , \tag{2.2.77}$$

and

$$\begin{aligned} b_n(t) &= b_n \, \mathrm{e}^{-\mathrm{i}v|k_n|t} , \\ b_n^\dagger(t) &= b_n^\dagger \, \mathrm{e}^{\mathrm{i}v|k_n|t} . \end{aligned} \tag{2.2.78}$$

From the above considerations, we conclude that the following expansion holds:

$$\begin{aligned} \phi(t, x) = {} & \frac{2\pi RN}{L} x + Q + \frac{\pi P v}{L} t \\ & + \frac{1}{2} \sum_{n \neq 0} \sqrt{\frac{1}{|n|}} \left(b_n \, \mathrm{e}^{\mathrm{i}k_n x - \mathrm{i}v|k_n|t} + b_n^\dagger \, \mathrm{e}^{-\mathrm{i}k_n x + \mathrm{i}v|k_n|t} \right) . \end{aligned} \tag{2.2.79}$$

Owing to the $2\pi R$-periodicity of $\phi(t, x)$, the value of P is quantized as

$$P = \frac{M}{R} , \tag{2.2.80}$$

with M being an integer. Writing $\Psi(Q)$ for the wave function of Q, owing to the condition $\Psi(Q + 2\pi R) = \Psi(Q)$, the eigenvalues of P are quantized as $\Psi(Q) \propto \mathrm{e}^{\mathrm{i}\frac{M}{R}Q}$ (M: integer). Then, considering $P = \frac{1}{\mathrm{i}}\partial/\partial Q$, equation (2.2.80) is derived.

Now, in (2.2.79) $|n|$ and $|k_n|$ appear, and therefore the left-moving and right-moving components are not totally independent. To see this point easily, we define from $b_n^\dagger$ and b_n with $n > 0$

$$\begin{aligned} b_n &= \mathrm{i}\sqrt{\frac{1}{n}} \bar{\alpha}_n , & b_{-n} &= \mathrm{i}\sqrt{\frac{1}{n}} \alpha_n , \\ b_n^\dagger &= -\mathrm{i}\sqrt{\frac{1}{n}} \bar{\alpha}_{-n} , & b_{-n}^\dagger &= -\mathrm{i}\sqrt{\frac{1}{n}} \alpha_{-n} . \end{aligned} \tag{2.2.81}$$

The commutation relation in these variables reads

$$\begin{aligned} [\alpha_m, \alpha_n] &= [\bar{\alpha}_m, \bar{\alpha}_n] = m\delta_{m+n,0} , \\ [\alpha_m, \bar{\alpha}_n] &= 0 \end{aligned} \tag{2.2.82}$$

and (2.2.79) becomes

$$\phi(t,x) = \frac{2\pi}{L}\left(RNx + \frac{M}{2R}vt\right) + Q + \frac{\mathrm{i}}{2}\sum_{n\neq 0}\frac{1}{n}\left[\alpha_n\,\mathrm{e}^{-\mathrm{i}k_n(x+vt)} + \bar{\alpha}_n\,\mathrm{e}^{-\mathrm{i}k_n(-x+vt)}\right]. \tag{2.2.83}$$

In the imaginary time formalism, z corresponds to the direction $x+vt \to x - \mathrm{i}v\tau = -\mathrm{i}(r_0 + \mathrm{i}r_1) = -\mathrm{i}\xi$, therefore α_n corresponds to the holomorphic, and $\bar{\alpha}_n$ to the anti-holomorphic components. Using the transformation (2.2.9), this correspondence leads to

$$\begin{aligned} \mathrm{e}^{\mathrm{i}\frac{2\pi}{L}(vt+x)} &\Rightarrow \mathrm{e}^{\mathrm{i}\frac{2\pi}{L}(-\mathrm{i}\xi)} = \mathrm{e}^{\frac{2\pi\xi}{L}} = z\,, \\ \mathrm{e}^{\mathrm{i}\frac{2\pi}{L}(vt-x)} &\Rightarrow \mathrm{e}^{\mathrm{i}\frac{2\pi}{L}(-\mathrm{i}\bar{\xi})} = \mathrm{e}^{\frac{2\pi\bar{\xi}}{L}} = \bar{z}\,. \end{aligned} \tag{2.2.84}$$

In these coordinates, (2.2.83) can be written as

$$\phi(z,\bar{z}) = \frac{1}{2}(\varphi(z) + \bar{\varphi}(\bar{z}))\,, \tag{2.2.85}$$

$$\begin{aligned} \varphi(z) &= Q - \mathrm{i}\alpha_0 \ln z + \mathrm{i}\sum_{n\neq 0}\frac{1}{n}z^{-n}\alpha_n\,, \\ \bar{\varphi}(\bar{z}) &= Q - \mathrm{i}\bar{\alpha}_0 \ln \bar{z} + \mathrm{i}\sum_{n\neq 0}\frac{1}{n}\bar{z}^{-n}\bar{\alpha}_n\,. \end{aligned} \tag{2.2.86}$$

Here, we defined

$$\alpha_0 = \frac{P}{2} + RN\,, \qquad \bar{\alpha}_0 = \frac{P}{2} - RN\,. \tag{2.2.87}$$

Equations (2.2.86) contain a logarithmic term, and taking the derivative with respect to z and $\bar{z}$ we can write the expansion in a compact manner as

$$\begin{aligned} \partial_z\varphi(z) &= -\mathrm{i}\sum_{n=-\infty}^{\infty} z^{-(n+1)}\alpha_n\,, \\ \partial_{\bar{z}}\bar{\varphi}(\bar{z}) &= -\mathrm{i}\sum_{n=-\infty}^{\infty} \bar{z}^{-(n+1)}\bar{\alpha}_n\,. \end{aligned} \tag{2.2.88}$$

Following (2.2.18), $T(z)$ becomes

$$\begin{aligned} T(z) &= -\frac{1}{2} :(\partial_z\varphi(z))^2: = \frac{1}{2}\sum_{n,m} z^{-(n+1)}z^{-(m+1)} :\alpha_n\alpha_m: \\ &= \sum_n z^{-(n+2)}L_n\,, \end{aligned} \tag{2.2.89}$$

and we conclude that

$$L_n = \frac{1}{2} \sum_m : \alpha_{m-n} \alpha_{-m} : . \tag{2.2.90}$$

In particular, we obtain

$$L_0 = \frac{1}{2} \alpha_0^2 + \sum_{n=1}^{\infty} \alpha_{-n} \alpha_n . \tag{2.2.91}$$

Recall that owing to the definition (2.2.81), $\alpha_n (n \geq 1)$ are annihilation operators. In the same way, we deduce

$$\bar{L}_0 = \frac{1}{2} \bar{\alpha}_0^2 + \sum_{n=1}^{\infty} \bar{\alpha}_{-n} \bar{\alpha}_n . \tag{2.2.92}$$

Now, we consider the Hamiltonian and the total momentum. Because these operators cause the time development and the spatial translation, the relation to $T_{\mu\nu}$ can be deduced easily. As demonstrated in Appendix B, the Hamiltonian and the total momentum are given by

$$H = v \int \frac{\mathrm{d}x}{2\pi} T_{00}(x,t) = v \int \frac{\mathrm{d}x}{2\pi} [T(\xi) + \bar{T}(\bar{\xi})] , \tag{2.2.93}$$

$$P = \mathrm{i}v \int \frac{\mathrm{d}x}{2\pi} T_{01}(x,t) = v \int \frac{\mathrm{d}x}{2\pi} [T(\xi) - \bar{T}(\bar{\xi})] . \tag{2.2.94}$$

Here, we set $\xi = \mathrm{i}vt + \mathrm{i}x$, $\bar{\xi} = \mathrm{i}vt - \mathrm{i}x$. When the conformal transformation (2.2.7) is applied and the ξ-plane is exchanged by the z-plane, using (2.2.49), we obtain

$$\tilde{T}(z) = \left(\frac{2\pi}{L}\right)^2 \left[T(z) z^2 - \frac{c}{24}\right] , \tag{2.2.95}$$

and therefore

$$\begin{aligned} \int \frac{\mathrm{d}x}{2\pi} T(\xi) &= \oint_c \frac{\mathrm{d}z}{2\pi \mathrm{i} z} \left(\frac{L}{2\pi}\right) \tilde{T}(z) \\ &= \frac{2\pi}{L} \oint_c \frac{\mathrm{d}z}{2\pi \mathrm{i} z} \left[T(z) z^2 - \frac{c}{24}\right] = \frac{2\pi}{L} \left(L_0 - \frac{c}{24}\right) . \end{aligned} \tag{2.2.96}$$

In the same manner,

$$\int \frac{\mathrm{d}x}{2\pi} \bar{T}(\bar{\xi}) = \frac{2\pi}{L} \left(\bar{L}_0 - \frac{c}{24}\right) \tag{2.2.97}$$

is obtained, and therefore, following (2.2.93) and (2.2.94), finally we get

$$H = \frac{2\pi v}{L}(L_0 + \bar{L}_0) - \frac{\pi c v}{6L}, \tag{2.2.98}$$

$$P = \frac{2\pi v}{L}(L_0 - \bar{L}_0). \tag{2.2.99}$$

These equations not only imply that the Hamiltonian and the total momentum operator can be expressed in terms of the conformal dimensions Δ and $\bar{\Delta}$, but also that the finite size energy correction $-(\pi c v)/(6L)$ depends only on the central charge and the velocity v. Conversely, from the energy and momentum eigenvalues of a finite system, the important quantities $\Delta, \bar{\Delta}, c$, being characteristic for the theory, can be read off.

Now, let us again apply the general theory to the special case of the action (2.2.3). Owing to (2.2.91), the eigenvalue of L_0 is given by

$$\Delta = \frac{1}{2}\left(\frac{M}{2R} + RN\right)^2 + \sum_{n<0} |n| N_n, \tag{2.2.100}$$

and, in the same manner, the eigenvalue of $\bar{L}_0$ is given by

$$\bar{\Delta} = \frac{1}{2}\left(\frac{M}{2R} - RN\right)^2 + \sum_{n>0} n N_n. \tag{2.2.101}$$

Here, N_n is a non-negative integer.

Now, we discuss the meaning of the first term in (2.2.100) and (2.2.101). In terms of the considerations made in Sect. 2.2, N_{R} and N_{L} are the numbers of right-moving and left-moving fermions, measured from the vacuum, respectively. Then, from (2.2.79), we obtain

$$\begin{aligned} N_{\mathrm{R}} + N_{\mathrm{L}} &= \int \frac{\mathrm{d}x}{2\pi} \partial_x \theta_+(x) = N, \\ N_{\mathrm{R}} - N_{\mathrm{L}} &= \int \frac{\mathrm{d}x}{2\pi} \partial_x \theta_-(x) = \int \frac{\mathrm{d}x}{2\pi} \frac{2}{\eta v} \partial_t \theta_+ = M. \end{aligned} \tag{2.2.102}$$

That is, N and M describe the variation in the right- and left-handed fermion branches. Furthermore, owing to (2.2.91)

$$[L_0, \alpha_0] = 0 \tag{2.2.103}$$

holds, and therefore both can be diagonalized simultaneously. That is, the first and second terms in (2.2.100) can be considered independently. The second term does not change the particle number in each branch and corresponds to the phonon excitations $\alpha_n (n \neq 0)$. A state that is a vacuum state for $\alpha_{n\neq 0}$ and has the eigenvalue $\beta = M/(2R) + RN$ of α_0 is given by

$$|\beta\rangle = \lim_{z \to 0} :\mathrm{e}^{\mathrm{i}\beta\varphi(z)}: |\mathrm{vac}\rangle. \tag{2.2.104}$$

In terms corresponding to (2.2.58), $\mathcal{O}(z) = :\mathrm{e}^{\mathrm{i}\beta\varphi(z)}:$ is a primary field with $\Delta = \beta^2/2$ and $\alpha_0|\beta\rangle = \beta|\beta\rangle$.

Let us confirm this explicitly. The first point can be seen by determining the OPE with $T(z)$

$$
\begin{aligned}
RT(z)\mathcal{O}(w) &= -\frac{1}{2} :(\partial_z\varphi(z))^2: :\mathrm{e}^{\mathrm{i}\beta\varphi(w)}: \\
&= -\frac{1}{2} :\partial_z\varphi(z)\partial_z\varphi(z)(:\mathrm{e}^{\mathrm{i}\beta\varphi(w)}:): \\
&\quad - :\partial_z\varphi(z)(:\partial_z\varphi(z)\,\mathrm{e}^{\mathrm{i}\beta\varphi(w)}:): \\
&= \frac{\beta^2/2}{(z-w)^2} :\mathrm{e}^{\mathrm{i}\beta\varphi(w)}: + \mathrm{i}\beta :\partial_z\varphi(z)\left(\frac{1}{z-w} :\mathrm{e}^{\mathrm{i}\beta\varphi(w)}:\right): \\
&= \frac{\beta^2/2}{(z-w)^2} :\mathrm{e}^{\mathrm{i}\beta\varphi(w)}: + \frac{1}{z-w}\partial_w :\mathrm{e}^{\mathrm{i}\beta\varphi(w)}: + \text{regular}\,.
\end{aligned}
\tag{2.2.105}
$$

Therefore, indeed, $:\mathrm{e}^{\mathrm{i}\beta\varphi(z)}:$ is a primary field with conformal dimension $\Delta = \beta^2/2$. Here, we used

$$
\langle\varphi(z)\varphi(w)\rangle = -\ln(z-w) \tag{2.2.106}
$$

and

$$
\begin{aligned}
\partial_z\varphi(z) :\mathrm{e}^{\mathrm{i}\beta\varphi(w)}: &= \sum_{n=0}^{\infty} \frac{(\mathrm{i}\beta)^n}{n!} \partial_z\varphi(z) :(\varphi(w))^n: \\
&= \sum_{n=0}^{\infty} \frac{(\mathrm{i}\beta)^n}{n!} n\partial_z\varphi(z) :\varphi(w)(\varphi(w))^{n-1}: + \text{regular} \\
&= \sum_{n=0}^{\infty} \frac{(\mathrm{i}\beta)^n}{(n-1)!}\frac{-1}{z-w} :(\varphi(w))^{n-1}: + \text{regular} \\
&= -\mathrm{i}\beta\frac{1}{z-w} :\mathrm{e}^{\mathrm{i}\beta\varphi(w)}: + \text{regular}\,.
\end{aligned}
\tag{2.2.107}
$$

Next, $\alpha_0|\beta\rangle = \beta|\beta\rangle$ can be confirmed using (2.2.88). Noting that α_0 can be expressed as

$$
\alpha_0 = \oint \frac{\mathrm{d}z}{2\pi\mathrm{i}}\, \mathrm{i}\partial_z\varphi(z)\,, \tag{2.2.108}
$$

we obtain using (2.2.107)

$$
\begin{aligned}
\alpha_0|\beta\rangle &= \lim_{w\to 0} \oint \frac{\mathrm{d}z}{2\pi\mathrm{i}}\, \mathrm{i}\partial_z\varphi(z) :\mathrm{e}^{\mathrm{i}\beta\varphi(w)}: |\mathrm{vac}\rangle \\
&= \lim_{w\to 0} \oint \frac{\mathrm{d}z}{2\pi\mathrm{i}} \frac{\beta}{z-w}|\beta\rangle = \beta|\beta\rangle\,.
\end{aligned}
\tag{2.2.109}
$$

In this manner, we have seen that the first term in (2.2.100) corresponds totally to a primary field. The second term is already expressed in a compact manner in terms of the bosonic occupation number N_n. To express this term in the language of Virasoro algebra, it is necessary to determine the primary fields expressed as polynomial in $\partial_z \phi(z)$, being responsible for the phonon-excitations. We already know that $\partial_z \phi(z)$ is such an example. It is quite difficult also to determine the other contributions.

This section will close by noting the results for the correlation function following conformal invariance. We consider the two-point function of primary fields $\mathcal{O}_i$ with conformal dimension $(\Delta_i, \bar{\Delta}_i)$

$$G^{(2)}(z_i, \bar{z}_i) = \langle \mathcal{O}_1(z_1, \bar{z}_1) \mathcal{O}_2(z_2, \bar{z}_2) \rangle \,. \tag{2.2.110}$$

The requirement that $G^{(2)}(z_i, \bar{z}_i)$ is invariant under $z \to z + \varepsilon(z)$, $\bar{z} \to \bar{z} + \bar{\varepsilon}(\bar{z})$ leads with (2.2.42) to

$$\begin{aligned}
&\big[(\varepsilon(z_1)\partial_{z_1} + \Delta_1 \partial_{z_1}\varepsilon(z_1)) + (\varepsilon(z_2)\partial_{z_2} + \Delta_2 \partial_{z_2}\varepsilon(z_2)) \\
&\quad + (\bar{\varepsilon}(\bar{z}_1)\partial_{\bar{z}_1} + \bar{\Delta}_1 \partial_{\bar{z}_1}\bar{\varepsilon}(\bar{z}_1)) + (\bar{\varepsilon}(\bar{z}_2)\partial_{\bar{z}_2} + \bar{\Delta}_2 \partial_{\bar{z}_2}\varepsilon(\bar{z}_2))\big] \\
&\quad \times G^{(2)}(z_i, \bar{z}_i) = 0 \,.
\end{aligned} \tag{2.2.111}$$

From the translation operation $\varepsilon(z) = 1$, $\bar{\varepsilon}(\bar{z}) = 1$, it follows from (2.2.111) that $G^{(2)}(z_i, \bar{z}_i)$ is a function of $z_{12} = z_1 - z_2$ and $\bar{z}_{12} = \bar{z}_1 - \bar{z}_2$ only. Furthermore, for the scaling transformation, $\varepsilon(z) = z$, $\bar{\varepsilon}(\bar{z}) = \bar{z}$, (2.2.111) becomes

$$\begin{aligned}
&[z_{12}\partial_{z_{12}} + (\Delta_1 + \Delta_2)] G^{(2)}(z_{12}, \bar{z}_{12}) = 0 \,, \\
&[\bar{z}_{12}\partial_{\bar{z}_{12}} + (\bar{\Delta}_1 + \bar{\Delta}_2)] G^{(2)}(z_{12}, \bar{z}_{12}) = 0 \,,
\end{aligned} \tag{2.2.112}$$

and therefore with C_{12} being a coefficient, we obtain

$$G^{(2)}(z_{12}, \bar{z}_{12}) = \frac{C_{12}}{z_{12}^{\Delta_1 + \Delta_2} \bar{z}_{12}^{\bar{\Delta}_1 + \bar{\Delta}_2}} \,. \tag{2.2.113}$$

Furthermore, when setting $\varepsilon(z) = z^2$, $\bar{\varepsilon}(\bar{z}) = \bar{z}^2$ and inserting the explicit form (2.2.113) in (2.2.111), we obtain

$$\begin{aligned}
&-(\Delta_1 + \Delta_2)(z_1 + z_2) + 2(\Delta_1 z_1 + \Delta_2 z_2) = 0 \,, \\
&-(\bar{\Delta}_1 + \bar{\Delta}_2)(\bar{z}_1 + \bar{z}_2) + 2(\bar{\Delta}_1 \bar{z}_1 + \bar{\Delta}_2 \bar{z}_2) = 0 \,,
\end{aligned} \tag{2.2.114}$$

and conclude that $\Delta_1 = \Delta_2 = \Delta$ and $\bar{\Delta}_1 = \bar{\Delta}_2 = \bar{\Delta}$. Otherwise, $G^{(2)}$ vanishes. Therefore, we finally obtain the form

$$G^{(2)}(z_{12}, \bar{z}_{12}) = \frac{C_{12}}{(z_{12})^{2\Delta} (\bar{z}_{12})^{2\bar{\Delta}}} \,. \tag{2.2.115}$$

That is, the critical exponents of the two-point correlation function of the primary fields are determined by the conformal dimensions.

We now apply these general considerations to an explicit example.

$$\mathcal{O}_1 = \mathcal{O}_2 = \mathrm{e}^{\mathrm{i}2\beta\phi(z,\bar{z})} \tag{2.2.116}$$

is a primary field with conformal dimension $(\Delta, \bar{\Delta}) = (\beta^2/2, \beta^2/2)$, as can be understood from the discussion following (2.2.104). Therefore, we obtain

$$\left\langle \mathrm{e}^{\mathrm{i}2\beta\phi(z,\bar{z})}\, \mathrm{e}^{-\mathrm{i}2\beta\phi(0,0)} \right\rangle = \frac{C_{12}}{|z|^{2\beta^2}} . \tag{2.2.117}$$

Because (2.2.115) is conformal invariant in the whole z-plane, it is regarded as the correlation function after applying the mapping (2.2.7). Following (2.2.44), the correlation function in the ξ-space is given by

$$\begin{aligned} G^{(2)}(\xi_i, \bar{\xi}_i) &= \left(\frac{\mathrm{d}z(\xi_1)}{\mathrm{d}\xi_1} \frac{\mathrm{d}z(\xi_2)}{\mathrm{d}\xi_2} \right)^{\Delta} \left(\frac{\mathrm{d}\bar{z}(\bar{\xi}_1)}{\mathrm{d}\bar{\xi}_1} \frac{\mathrm{d}\bar{z}(\bar{\xi}_2)}{\mathrm{d}\bar{\xi}_2} \right)^{\bar{\Delta}} \\ &\quad \times G^{(2)}(z(\xi_1) - z(\xi_2), \bar{z}(\xi_1) - \bar{z}(\xi_2)) \\ &= C_{12} \left\{ \frac{\pi/L}{\sinh[\pi(\xi_1 - \xi_2)/L]} \right\}^{2\Delta} \left\{ \frac{\pi/L}{\sinh[\pi(\bar{\xi}_1 - \bar{\xi}_2)/L]} \right\}^{2\bar{\Delta}} . \end{aligned} \tag{2.2.118}$$

This is the correlation function of the finite system, having the striking property that it can be deduced from conformal invariance. Using the expansion

$$\begin{aligned} \frac{1}{\sinh\left[\dfrac{\pi(\xi_1 - \xi_2)}{L} \right]} &= 2\,\mathrm{e}^{-\pi(\xi_1 - \xi_2)/L} \left[1 - \mathrm{e}^{-2\pi(\xi_1 - \xi_2)/L} \right]^{-1} \\ &= 2\,\mathrm{e}^{-\pi(\xi_1 - \xi_2)/L} \sum_{n=0}^{\infty} \mathrm{e}^{-2\pi n(\xi_1 - \xi_2)/L} , \end{aligned} \tag{2.2.119}$$

(2.2.118) can be expressed as

$$G^{(2)}(\xi_i, \bar{\xi}_1) = \sum_{n,m} a_{nm}\, \mathrm{e}^{-2\pi(\Delta + n)(\xi_1 - \xi_2)/L}\, \mathrm{e}^{-2\pi(\bar{\Delta} + m)(\bar{\xi}_1 - \bar{\xi}_2)/L} \tag{2.2.120}$$

with $a_{n,m}$ being some coefficients. On the other hand, let $|q\rangle$ be the set of states with energy eigenvalue E_q and momentum eigenvalue P_q. Because they built up a basis, with $|0\rangle$ being the ground state, $G^2(\xi_i, \bar{\xi}_i)$ can be expressed as

$$\begin{aligned} G^{(2)}(\xi_1, \bar{\xi}_2) &= \sum_q \langle 0|\mathcal{O}_1(0,0)|q\rangle \langle q|\mathcal{O}_2(0,0)|0\rangle \\ &\quad \times \mathrm{e}^{-(E_q - E_0)(\tau_1 - \tau_2) - \mathrm{i}P_q(x_1 - x_2)/v} \end{aligned} \tag{2.2.121}$$

with $\tau_1 - \tau_2 = (\xi_1 - \xi_2 + \bar{\xi}_1 - \bar{\xi}_2)/2v$ and $\mathrm{i}(x_1 - x_2) = (\xi_1 - \xi_2 - \bar{\xi}_1 + \bar{\xi}_2)/2$. Comparing (2.2.120) and (2.2.121), we obtain the relations

$$v^{-1}(E_q + P_q) = \frac{4\pi}{L}(\Delta + n)\,,$$
$$v^{-1}(E_q - P_q) = \frac{4\pi}{L}(\bar{\Delta} + m)\,. \tag{2.2.122}$$

With n_1 and n_2 being integers, we conclude that

$$E_q = \frac{2\pi v}{L}(\Delta + \bar{\Delta} + n_1)\,,$$
$$P_q = \frac{2\pi v}{L}(\Delta - \bar{\Delta} + n_2)\,. \tag{2.2.123}$$

Recalling that Δ and $\bar{\Delta}$ are the eigenvalues of L_0 and $\bar{L}_0$, we conclude that we have reproduced the conformal tower (2.2.98) and (2.2.99).

2.3 The Non-linear Sigma Model: Effective Theory of Quantum Anti-ferromagnets

In this section, another, complementary approach to the one-dimensional quantum anti-ferromagnet introduced in the previous section will be presented. This method is not restricted to one dimension, but can be applied in any dimension, and provides a good semiclassical approximation when the spin S is large.

The principal idea is to construct an effective action for the 'slowly varying degrees of freedom'. The Hamiltonian is given by

$$H = \sum_{i,j} J_{ij} \boldsymbol{S}_i \cdot \boldsymbol{S}_j \tag{2.3.1}$$

and the partition function, expressed as a path integral, reads

$$Z = \int \prod_i \mathcal{D}\boldsymbol{S}_i(\tau)\, \mathrm{e}^{-A} \tag{2.3.2}$$

with action A given by

$$A = \mathrm{i}S \sum_i \omega(\{\boldsymbol{S}_i(\tau)\}) + \int_0^\beta \mathrm{d}\tau \sum_{ij} J_{ij} \boldsymbol{S}_i(\tau) \cdot \boldsymbol{S}_j(\tau)\,. \tag{2.3.3}$$

In the following, we consider a d-dimensional lattice with lattice constant a, and $J_{ij} = J > 0$ only acts between nearest-neighbour states. The first term on the right-hand side is called Berryphase. When $\boldsymbol{S}_i(\tau)$ moves around starting at $\tau = 0$ until $\tau = \beta$, $\omega(\{\boldsymbol{S}_i(\tau)\})$ is given by the solid angle subtended by the trajectory of $\boldsymbol{n}_i(\tau) = \boldsymbol{S}_i(\tau)/S$, on the surface of the unit sphere. In terms of the polar coordinates $n = (\sin\theta\,\cos\varphi,\ \sin\theta\,\sin\varphi,\ \cos\theta)$,

$$\omega(\{\boldsymbol{S}(\tau)\}) = \int_0^\beta \mathrm{d}\tau\,(1 - \cos\theta(\tau))\dot{\varphi}(\tau)\,. \tag{2.3.4}$$

Equations (2.3.2) and (2.3.3) are exact equations, and in the following we will derive the effective long-wavelength, low-energy action in the continuum limit. First, we will specify what is meant by 'slowly varying degrees of freedom'. Two principal ideas will show us the way to do so.

The first point is the conservation laws. In the present case, the total spin $\boldsymbol{S}_{\text{tot}} = \sum_i \boldsymbol{S}_i$ commutes with the Hamiltonian (2.3.1), and therefore $\boldsymbol{S}_{\text{tot}}$ is a conserved quantity and time-independent. That is, after Fourier transformation, $\boldsymbol{S}(\boldsymbol{q}=0)$ is time-independent, and therefore we expect that for small $|\boldsymbol{q}|$, ($|\boldsymbol{q}| \ll \pi/a$), $\boldsymbol{S}(\boldsymbol{q})$ is only weakly time-dependent. Here, a is the lattice constant.

The second point is symmetry breaking and the corresponding Goldstone mode. Because we consider J_{ij} for the anti-ferromagnetic case,

$$\langle \boldsymbol{S}_i \rangle = (-1)^i \boldsymbol{M}_S \tag{2.3.5}$$

is expected to describe the anti-ferromagnetic long-range order. The sign is given by

$$(-1)^i = \mathrm{e}^{\mathrm{i}\boldsymbol{Q}\cdot\boldsymbol{R}_i}, \tag{2.3.6}$$

where $\boldsymbol{Q}$ is defined as having the value π/a in all components, and $\boldsymbol{R}_i$ is the position coordinate of site i. When the direction of $\boldsymbol{M}_S$ is changed simultaneously at all sites, the energy is degenerate. Variations of $\boldsymbol{M}_S$ in the large wavelength limit have a frequency close to zero (Goldstone mode). This corresponds to $|\boldsymbol{q}|$ vectors in $\boldsymbol{S}(\boldsymbol{q})$ nearby $\boldsymbol{Q}$. Therefore, also in this limit, we expect only slowly varying modes.

Notice that these considerations can also be applied in the case when no long-range order is present. That is, for the case when strong anti-ferromagnetic correlation over long distances is present, approximately long-range order can be assumed, and $\boldsymbol{S}(\boldsymbol{q})$ for $|\boldsymbol{q}-\boldsymbol{Q}| \ll \pi/a$ is slowly changing.

From the above considerations, we write with $\boldsymbol{\Omega}(\boldsymbol{R}_i)$ and $\boldsymbol{L}(\boldsymbol{R}_i)$ in the case of slow time and space dependence

$$\boldsymbol{S}_i(\tau) = (-1)^i S\boldsymbol{\Omega}(\boldsymbol{R}_i) + a\boldsymbol{L}(\boldsymbol{R}_i). \tag{2.3.7}$$

Here, $\boldsymbol{\Omega}$ is the unit vector ($|\boldsymbol{\Omega}| = 1$), and in order that the normalization condition $|\boldsymbol{S}_i(\tau)|^2 = S^2$ is satisfied up to order a,

$$\boldsymbol{\Omega}(\boldsymbol{R}_i)\cdot\boldsymbol{L}(\boldsymbol{R}_i) = 0 \tag{2.3.8}$$

must hold. In (2.3.7), $\boldsymbol{L}(\boldsymbol{R}_i)$ is multiplied by a for the following reason. Assuming perfect anti-ferromagnetic ordering (for example, $\boldsymbol{\Omega} = \boldsymbol{e}_z$), no effective magnetization occurs, and because the effective magnetization is related to the space derivative of $\boldsymbol{\Omega}$, as can be seen in (2.3.9), the continuum limit becomes consistent when the factor a is introduced.

We insert (2.3.7) into (2.3.3). With $\langle ij \rangle$ being the nearest-neighbouring pair, the Hamiltonian is given by

$$
\begin{aligned}
H &= J \sum_{\langle ij \rangle} \boldsymbol{S}_i \cdot \boldsymbol{S}_j \\
&= J \sum_{\langle ij \rangle} \left[(-1)^i S \boldsymbol{\Omega}(\boldsymbol{R}_i) + a \boldsymbol{L}(\boldsymbol{R}_i) \right] \left[(-1)^j S \boldsymbol{\Omega}(\boldsymbol{R}_j) + a \boldsymbol{L}(\boldsymbol{R}_j) \right] \\
&= -JS^2 \sum_{i,\alpha} \boldsymbol{\Omega}(\boldsymbol{R}_i) \cdot \boldsymbol{\Omega}(\boldsymbol{R}_i + a\boldsymbol{e}_\alpha) \\
&\quad + JSa \sum_{i,\alpha} (-1)^i \{ \boldsymbol{\Omega}(\boldsymbol{R}_i) \cdot \boldsymbol{L}(\boldsymbol{R}_i + a\boldsymbol{e}_\alpha) + \boldsymbol{\Omega}(\boldsymbol{R}_i) \cdot \boldsymbol{L}(\boldsymbol{R}_i - a\boldsymbol{e}_\alpha) \} \\
&\quad + Ja^2 \sum_{i,\alpha} \boldsymbol{L}(\boldsymbol{R}_i) \cdot \boldsymbol{L}(\boldsymbol{R}_i + a\boldsymbol{e}_\alpha) \\
&= -JS^2 \sum_{i,\alpha} \left\{ 1 + a \boldsymbol{\Omega}(\boldsymbol{R}_i) \cdot \frac{\partial \boldsymbol{\Omega}(\boldsymbol{R}_i)}{\partial R_i^\alpha} + \frac{1}{2} a^2 \boldsymbol{\Omega}(\boldsymbol{R}_i) \cdot \frac{\partial^2 \boldsymbol{\Omega}(\boldsymbol{R}_i)}{\partial R_i^{\alpha 2}} \right\} \\
&\quad + 2JSa^2 \sum_{i,\alpha} (-1)^i \boldsymbol{\Omega}(\boldsymbol{R}_i) \cdot \boldsymbol{L}(\boldsymbol{R}_i) + Ja^2 d \sum_i |\boldsymbol{L}(\boldsymbol{R}_i)|^2 + O(a^3) .
\end{aligned}
\tag{2.3.9}
$$

Here, α stands for the directions x, y, etc. Because of $|\boldsymbol{\Omega}|^2 = 1$, it follows that $\boldsymbol{\Omega} \cdot \partial \boldsymbol{\Omega} / \partial R^a = 0$. Owing to (2.3.8), (2.3.9) has in the continuum limit the following integral representation

$$
\begin{aligned}
H &= -JS^2 \sum_{i,\alpha} \left\{ 1 + \frac{1}{2} a^2 \boldsymbol{\Omega}(\boldsymbol{R}_i) \cdot \frac{\partial^2 \boldsymbol{\Omega}(\boldsymbol{R}_i)}{\partial R_i^{\alpha 2}} \right\} \\
&\quad + Jda^2 \sum_i |\boldsymbol{L}(\boldsymbol{R}_j)|^2 + O(a^3) \\
&\cong -JS^2 \sum_{i,\alpha} 1 + \int \mathrm{d}^d \boldsymbol{R} \left\{ \frac{JS^2 a^{2-d}}{2} |\nabla \boldsymbol{\Omega}(\boldsymbol{R})|^2 + Jda^{2-d} \boldsymbol{L}(\boldsymbol{R})^2 \right\} .
\end{aligned}
\tag{2.3.10}
$$

Next, we consider the Berry phase, and with $\boldsymbol{n}_i(\tau) = \boldsymbol{S}_i(\tau)/S$, we obtain

$$
\begin{aligned}
S \sum_i \omega(\{\boldsymbol{n}_i(\tau)\}) &= S \sum_i \omega \left(\left\{ (-1)^i \boldsymbol{\Omega}(\boldsymbol{R}_i, \tau) + \frac{a}{S} \boldsymbol{L}(\boldsymbol{R}_i, \tau) \right\} \right) \\
&\cong S \sum_i \left\{ \omega(\{(-1)^i \boldsymbol{\Omega}(\boldsymbol{R}_i)\}) \right. \\
&\quad \left. + \int_0^\beta \mathrm{d}\tau \frac{\delta \omega(\{\boldsymbol{n}(\tau)\})}{\delta \boldsymbol{n}(\tau)} \bigg|_{\boldsymbol{n}(\tau) = (-1)^i \boldsymbol{Q}(\boldsymbol{R}_i, \tau)} \cdot \frac{a}{S} \boldsymbol{L}(\boldsymbol{R}_i, \tau) \right\} .
\end{aligned}
\tag{2.3.11}
$$

Using $-\boldsymbol{n} = (\sin(\pi - \theta)\cos(\varphi + \pi), \sin(\pi - \theta)\sin(\varphi + \pi), \cos(\pi - \theta))$, we obtain

$$
\begin{aligned}
S\omega(\{-\boldsymbol{n}(\tau)\}) &= S \int_0^\beta \mathrm{d}\tau \, (1 + \cos\theta(\tau)) \dot{\varphi}(\tau) \\
&= -S\omega(\{\boldsymbol{n}(\tau)\}) + 2S[\varphi(\beta) - \varphi(0)]
\end{aligned}
\tag{2.3.12}
$$

and because $2S[\varphi(\beta) - \varphi(0)]$ is an integer times 2π, we can conclude that

$$\mathrm{e}^{-\mathrm{i}S\omega(\{-\boldsymbol{n}(\tau)\})} = \mathrm{e}^{\mathrm{i}S\omega(\{\boldsymbol{n}(\tau)\})}\,. \tag{2.3.13}$$

We rewrite the first term of the last equation on the right-hand side as

$$S\sum_i (-1)^i \omega(\{\boldsymbol{\Omega}(\boldsymbol{R}_i)\})\,. \tag{2.3.14}$$

Furthermore, owing to

$$\frac{\delta\omega(\{\boldsymbol{n}(\tau)\})}{\delta\boldsymbol{n}(\tau)} = \frac{\partial\boldsymbol{n}(\tau)}{\partial\tau} \times \boldsymbol{n}(\tau)\,, \tag{2.3.15}$$

the second term on the right-hand side of (2.3.11) can be expressed as

$$\sum_i \int_0^\beta \mathrm{d}\tau\, a\boldsymbol{L}(\boldsymbol{R}_i,\tau)\cdot\frac{\partial\boldsymbol{\Omega}(\boldsymbol{R}_i,\tau)}{\partial\tau} \times \boldsymbol{\Omega}(\boldsymbol{R}_i,\tau)\,. \tag{2.3.16}$$

Notice that the factor $(-1)^i$ appears twice and cancels when $\boldsymbol{n} = (-1)^i\boldsymbol{\Omega}$ is inserted.

At this stage, the effective action can be expressed as

$$\begin{aligned} A_{\mathrm{eff}} &= \mathrm{i}S\sum_i (-1)^i \omega(\{\boldsymbol{\Omega}(\boldsymbol{R}_i,\tau)\}) \\ &+ \int_0^\beta \mathrm{d}\tau \int \mathrm{d}^d\boldsymbol{R} \left\{ \frac{JS^2a^{2-d}}{2} |\nabla\boldsymbol{\Omega}(\boldsymbol{R},\tau)|^2 + Jda^{2-d}|\boldsymbol{L}(\boldsymbol{R},\tau)|^2 \right. \\ &\left. + \mathrm{i}a^{1-d}\boldsymbol{L}(\boldsymbol{R},\tau)\cdot\left(\frac{\partial\boldsymbol{\Omega}(\boldsymbol{R},\tau)}{\partial\tau} \times \boldsymbol{\Omega}(\boldsymbol{R},\tau)\right)\right\}. \end{aligned} \tag{2.3.17}$$

Notice that the homogeneous component $\boldsymbol{L}$ and the staggered component $\boldsymbol{\Omega}$ satisfy a canonical commutation relation corresponding to momentum and position coordinate, respectively. That is, because the integral of $\boldsymbol{L}$ is Gaussian, this suggests the relation

$$\boldsymbol{L} \sim \frac{-\mathrm{i}}{2Jda}\left(\frac{\partial\boldsymbol{\Omega}}{\partial\tau} \times \boldsymbol{\Omega}\right). \tag{2.3.18}$$

Although the integration over $\boldsymbol{L}$ must be performed under the constraint (2.3.8), because (2.3.18) is already orthogonal to $\boldsymbol{\Omega}$, finally it is sufficient to complete the square as usual to obtain the effective action expressed in $\boldsymbol{\Omega}$ only

$$\begin{aligned} A_{\mathrm{eff}} &= \mathrm{i}S\sum_i (-1)^i \omega(\{\boldsymbol{\Omega}(\boldsymbol{R}_i)\}) \\ &+ \int_0^\beta \mathrm{d}\tau \int \mathrm{d}^d\boldsymbol{R} \left\{ \frac{JS^2a^{2-d}}{2} |\nabla\boldsymbol{\Omega}(\boldsymbol{R},\tau)|^2 + \frac{a^{-d}}{4Jd}\left(\frac{\partial\boldsymbol{\Omega}(\boldsymbol{R},\tau)}{\partial\tau}\right)^2 \right\}. \end{aligned} \tag{2.3.19}$$

This is the so-called non-linear sigma model.

Here, we introduced the spin-wave velocity v and the coupling constant g by

$$c = \sqrt{2d}\, JSa\,, \tag{2.3.20a}$$

$$g = \frac{2\sqrt{2d}}{S} a^{d-1}\,. \tag{2.3.20b}$$

Defining $x_0 = c\tau$, $x_\alpha = R_\alpha$ as the $d+1$ dimensional coordinate x_μ, (2.3.19), becomes

$$A_{\text{eff}} = \mathrm{i}S\sum_i (-1)^i \omega(\{\boldsymbol{\Omega}(\boldsymbol{R}_i)\}) + \frac{1}{g}\int_0^{c\beta} \mathrm{d}x_0 \int \mathrm{d}^d\boldsymbol{x}\, (\partial_\mu \boldsymbol{\Omega})^2\,. \tag{2.3.21}$$

We now discuss the Berry phase appearing in the first term on the right-hand side of (2.3.21) in more detail. We start by considering the one-dimensional case. Imposing periodic boundary conditions assuming an even number of sites $\boldsymbol{\Omega}(\boldsymbol{R}_{2N+1}) = \boldsymbol{\Omega}(\boldsymbol{R}_1)$, we obtain

$$\begin{aligned} &S\sum_{i=1}^{2N} (-1)^i \omega(\{\boldsymbol{\Omega}(\mathrm{i}a)\}) \\ &\quad = S\sum_{k=1}^{N} \left[\omega(\{\boldsymbol{\Omega}(2ka)\}) - \omega(\{\boldsymbol{\Omega}((2k-1)a)\})\right] \\ &\quad \cong \frac{S}{2}\int_0^\beta \mathrm{d}\tau \int \mathrm{d}x\, \frac{\delta\omega(\{\boldsymbol{\Omega}(x,\tau)\})}{\delta\boldsymbol{\Omega}(x,\tau)} \frac{\partial\boldsymbol{\Omega}(x,\tau)}{\partial x} \\ &\quad = \frac{S}{2}\int_0^{\beta c} \mathrm{d}x_0 \int_0^{L=2Na} \mathrm{d}x_1\, \frac{\partial\boldsymbol{\Omega}(x)}{\partial x_0} \times \boldsymbol{\Omega}(x) \cdot \frac{\partial\boldsymbol{\Omega}(x)}{\partial x_1}\,. \end{aligned} \tag{2.3.22}$$

The factor 1/2 appeared because two sites are treated together.

The integral that we finally obtained has the following topological meaning. Considering the mapping of the two-dimensional sphere onto the unit sphere

$$(x_0, x_1) \mapsto \boldsymbol{\Omega}(x_0, x_1)\,, \tag{2.3.23}$$

the infinitesimal surface element translates into $\mathrm{d}\boldsymbol{\mathcal{A}}$:

$$\mathrm{d}\boldsymbol{\mathcal{A}} = \left(\frac{\partial\boldsymbol{\Omega}}{\partial x_0} \times \frac{\partial\boldsymbol{\Omega}}{\partial x_1}\right) \mathrm{d}x_0\, \mathrm{d}x_1\,. \tag{2.3.24}$$

Because this vector is parallel to $\boldsymbol{\Omega}$, the infinitesimal surface element, taking care of the correct sign convention, is given by

$$\mathrm{d}\mathcal{A} = \mathrm{d}\boldsymbol{\mathcal{A}} \cdot \boldsymbol{\Omega}\,.$$

The integration in the region $[0,\beta] \times [0,L]$ in the (x_0, x_1) space translates through the projection (2.3.23) into an integral on the unit sphere.

$\boldsymbol{\Omega}(x)$ is periodic both in the x_0 and x_1 directions, and assuming that $\boldsymbol{\Omega}$ is pointing in the same direction on the boundary of $[0,\beta]\times[0,L]$, $\boldsymbol{\Omega}$ defines a mapping from the unit sphere to the unit sphere. The number

$$Q=\frac{1}{4\pi}\int \mathrm{d}x_0\int \mathrm{d}x_1\left(\frac{\partial\boldsymbol{\Omega}}{\partial x_0}\times\frac{\partial\boldsymbol{\Omega}}{\partial x_1}\right)\cdot\boldsymbol{\Omega} \tag{2.3.25}$$

counts how often the unit sphere is wrapped by this projection. Q is called the Skyrmion number. We conclude that the first term on the right-hand side in (2.3.21), the Berry phase, is given by

$$\mathrm{e}^{-\mathrm{i}2\pi SQ}\,. \tag{2.3.26}$$

In the case when S is an integer, because Q is also an integer, $e^{-\mathrm{i}2\pi SQ}=1$, and the Berry phase can be ignored. However, for a half-odd-integer spin, (2.3.26) becomes

$$(-1)^Q \tag{2.3.27}$$

and depending on whether Q is odd or even, the sign does change.

That the low-energy properties of the one-dimensional quantum spin system are drastically different between $S=$ integer and $S=$ half-odd-integer has been discovered by Haldene. Before entering the quantitative analysis, we will give a qualitative picture. The path integral (2.3.2) can be interpreted as integration over all possible spin fluctuation configurations. In particular, for the case when the spin S is an integer where the Berry phase in (2.3.21) can be ignored, A_{eff} is positive. Because $e^{-A_{\text{eff}}}$ is positive all the time, all the quantum fluctuations in the path integral contribute with the same sign. However, for half-odd-integer spin, owing to the factor (2.3.27), the effect of the quantum fluctuations is suppressed owing to destructive interference caused by the alternating sign.

Therefore, the staggered component $\boldsymbol{\Omega}$ shows strong quantum effects for integer spin; on the other hand, it behaves more or less classically in the half-odd-integer spin case. As a result, the correlation function of $\boldsymbol{\Omega}$ shows longer-range correlation in the half-odd-integer spin case. With ξ being the correlation length, and m the gap in the excitation spectrum, the relation $\xi\propto m^{-1}$ holds. For integer spin, a finite m (Haldene gap) and a finite ξ is obtained; on the other hand, for half-odd-integer spin, m vanishes (gapless), ξ becomes infinite ($\xi=\infty$), and the correlation function shows a power-law decay. The case when $S=1/2$, which has been examined intensively in the last chapter, is an example of the latter case.

In such a way, owing to interference by the Berry phase, the systems becomes more classical. Considering the quantum mechanical path integral

$$\int \mathcal{D}x(t)\,\mathrm{e}^{\mathrm{i}S(\{x(t)\})/\hbar}\,, \tag{2.3.28}$$

only the classical path satisfying $\delta S = 0$ contributes in the limit $\hbar \to 0$. Also in the imaginary time formalism considered here, the Berry phase still remains purely imaginary, causing quantum interference.

Next, we consider the case of two and three dimensions. Regarding the two-, and three-dimensional lattice as an assembly of many one-dimensional chains, in every chain strong anti-ferromagnetic correlation is present. Labelling the chains with i, every chain has a Skyrmion number Q_i and a factor $(-1)^i$. When taking the sum over i, the Berry phase is expected to cancel out. However, this conclusion is only valid in the continuum limit for the slowly varying modes $\boldsymbol{\Omega}$. For the case when discontinuities or singularities are present (this is allowed for the original $\boldsymbol{\Omega}$ defined on the lattice points), the Berry phase becomes important. However, this case corresponds to a spin liquid where the anti-ferromagnetic long-range order disappeared owing to the quantum fluctuations. For the two- and three-dimensional Heisenberg model at zero temperature, as described in what follows, long-range order is present and the Berry phase in (2.3.21) can be ignored.

From the above considerations, we conclude that the Berry phase is only important for the half-odd-integer spin system in one dimension (for example $S = 1/2$). The analysis of the non-linear sigma model with a relevant Berry phase is a difficult task. However, we can expect that the low-energy properties of the effective action (2.3.21) are different depending on whether the spin S is integer or half-odd-integer, and qualitatively similar for each group. This assumption leads to the conclusion that in the half-odd-integer spin case, as has been investigated intensively for $S = 1/2$ in the foregoing chapter, a gapless excitation spectrum arises. Here, we will accept this conclusion without further investigations. Now, we consider the integer spin S case in a single dimension and in higher dimensions, where the Berry phase can be ignored. For simplicity, we set $c = a = 1$.

The partition function in the path integral representation is given by

$$Z = \int \mathcal{D}\boldsymbol{\Omega}(x) \prod_x \delta(\boldsymbol{\Omega}^2(x) - 1) \exp\left[-\frac{1}{g} \int \mathrm{d}^{d+1}x \, (\partial_\mu \boldsymbol{\Omega}(x))^2 \right] . \tag{2.3.29}$$

The action is quadratic in $\boldsymbol{\Omega}$, and the non-linearity arises owing to the constraint that has been introduced as a δ-function in the path integral

$$\boldsymbol{\Omega}^2(x) = 1 . \tag{2.3.30}$$

This condition can also be expressed as

$$\prod_x \delta(\boldsymbol{\Omega}^2(x) - 1) = \int \mathcal{D}\lambda(x) \exp\left[-\int \lambda(x)(\boldsymbol{\Omega}^2(x) - 1) \, \mathrm{d}^{d+1}x \right] . \tag{2.3.31}$$

Here, the integration of every $\lambda(x)$ runs over the region $[-\mathrm{i}\infty, +\mathrm{i}\infty]$. Constant factors such as 2π have been ignored.

The partition function Z then becomes

$$Z = \int \mathcal{D}\lambda(x)\mathcal{D}\boldsymbol{\Omega}(x) \exp\left[- \int \mathrm{d}^{d+1}x \left\{ \frac{1}{g}(\partial_\mu \boldsymbol{\Omega}(x))^2 + \lambda(x)(\boldsymbol{\Omega}^2(x) - 1) \right\} \right] . \tag{2.3.32}$$

Because the $\boldsymbol{\Omega}$-integration can now be performed without constraints, we obtain the result

$$Z = \int \mathcal{D}\lambda(x) \exp\left[\int \lambda(x)\, \mathrm{d}^{d+1}x \; - 3\,\mathrm{Tr}\ln\left\{ \frac{1}{g}\partial_\mu^2 + \lambda(x) \right\} \Big/ 2 \right] . \tag{2.3.33}$$

Here, the factor 3 is the number of components of $\boldsymbol{\Omega}$ (being a real field). The path integration of $\lambda(x)$ is difficult, because the term with Tr ln is a non-linear functional of λ.

Here, we apply the saddle point method to the path integral. Defining $-A_{\text{eff}}$ as the exponent in (2.3.33), the saddle point solution is obtained by the equation

$$\delta A_{\text{eff}}(\{\lambda(x)\}) = 0 . \tag{2.3.34}$$

In particular, for $\lambda(x) = \lambda = \text{const}$, the solution (at zero temperature) is given by

$$1 = \frac{3}{2}\,\mathrm{Tr}\, \frac{1}{\frac{1}{g}\partial_\mu^2 + \lambda} . \tag{2.3.35}$$

Here, Tr can be expressed in the energy-momentum basis as

$$\mathrm{Tr}\,\mathcal{O} \equiv \int \frac{\mathrm{d}\omega}{2\pi} \int \frac{\mathrm{d}^d \boldsymbol{k}}{(2\pi)^d} \langle \omega, \boldsymbol{k} | \mathcal{O} | \omega, \boldsymbol{k} \rangle . \tag{2.3.36}$$

Therefore, (2.3.35) becomes at zero temperature

$$\frac{2}{3} = \int_{|\boldsymbol{k}|<\Lambda} \frac{\mathrm{d}^{d+1}\boldsymbol{k}}{(2\pi)^{d+1}} \frac{1}{(k^2/g) + \lambda} , \tag{2.3.37}$$

where we introduced the cut-off $\Lambda \propto \pi/a$ of the wave number.

Now, we consider (2.3.37) in each dimension. First, for $d = 1$, logarithmic divergence occurs

$$\int_0^\Lambda \frac{\mathrm{d}^2 k}{(2\pi)^2} \frac{g}{k^2 + g\lambda} = \frac{g}{4\pi} \ln \frac{\Lambda^2 + g\lambda}{g\lambda} . \tag{2.3.38}$$

Here, $g\lambda$ is related to the gap in the spin excitation spectrum m by $g\lambda = m^2$. This is owing to the fact that when setting $\lambda = \text{const} = m^2/g$, the action in terms of $\boldsymbol{\Omega}$ becomes

$$\frac{1}{g} \sum_{\omega, \boldsymbol{k}} (\omega^2 + \boldsymbol{k}^2 + m^2) \boldsymbol{\Omega}(\omega, \boldsymbol{k}) \cdot \boldsymbol{\Omega}(-\omega, -\boldsymbol{k}) , \tag{2.3.39}$$

where we can read off the dispersion relation of the spin waves

$$\omega(\boldsymbol{k}) = \sqrt{\boldsymbol{k}^2 + m^2}\,. \tag{2.3.40}$$

Equation (2.3.38) determines this gap. The right-hand side is a decreasing function that for $m \to 0$ diverges logarithmically as $g/2\pi \ln \Lambda/m$. Therefore, for arbitrarily small g, a finite m

$$m \sim \Lambda\,\mathrm{e}^{-4\pi/3g} \tag{2.3.41}$$

arises.

This consideration is similar to the spin wavelength analysis of the long-range order in low-dimensional anti-ferromagnets. In one dimension, the divergence of the spin wave fluctuation corresponds to setting $m = 0$ in (2.3.38), and owing to a finite gap, these fluctuations can be suppressed. Then, the system becomes a quantum spin liquid (Haldene state) with finite correlation length. m is called the Haldene gap, being the most important parameter characterizing integer spin one-dimensional anti-ferromagnets.

In two and three dimensions, different from the one-dimensional case, no infrared divergence occurs. For $d = 2$, the integral is given by

$$\int_0^{\Lambda} \frac{\mathrm{d}^2 k}{(2\pi)^3} \frac{g}{k^2 + m^2} = \frac{g}{2\pi^2} \int_0^{\Lambda} \mathrm{d}k\, \frac{k^2}{k^2 + m^2}\,. \tag{2.3.42}$$

For $m = 0$, the right-hand side becomes $g\Lambda/2\pi^2$, and inserting this result in the equation (2.3.20b) for g, with $\Lambda \propto \pi/a$, we obtain

$$\frac{1}{2\pi^2}\frac{4}{S} a\Lambda \sim \frac{2}{\pi}\frac{1}{S}\,. \tag{2.3.43}$$

This is a dimensionless number depending only on S. We conclude that there exists a critical value S_{c}. For $S < S_{\mathrm{c}}$, (2.3.37) has a solution with finite m; however, for $S > S_{\mathrm{c}}$, the right-hand side of (2.3.3) becomes smaller than 2/3.

The former case is essentially similar to the Haldene state. The latter is an almost classical high-spin state, where long-range ferro-magnetic order should arise. Indeed, similar to the discussion of Bose condensation, the difference between the integral (2.3.37) and 2/3 gives rise to the condensed part of $\boldsymbol{\Omega}$, corresponding to long-range order. So, what might be the value of S_{c}? We cannot expect to find the exact value just by considering the effective theory in the continuum limit. It is now believed that the upper boundary for S_{c} is less than 1/2, and in the physically relevant case of $S \geq 1/2$ the antiferromagnetic long-range order occurs at zero temperature. This will be pointed out in what follows.

Because experiments are performed at finite temperature, the temperature dependence of physical quantities contains important information. Since long-range order in two dimensions at finite temperature does not exist (Mermin–Wagner theorem), the behaviour of the correlation length ξ in dependence on temperature is crucial. ξ can be observed through the scattering of neutrons. In the case of finite temperature, in the k integral (2.3.36) the contribution $k_0 = \omega$ in (2.3.36) is discretized in terms of Matsubara frequencies $\omega_n = 2\pi T$.

$$\operatorname{Tr} \mathcal{O} = \frac{1}{\beta} \sum_{\omega_n} \int \frac{\mathrm{d}^d \boldsymbol{k}}{(2\pi)^d} \langle \omega_n, \boldsymbol{k} | \mathcal{O} | \omega_n, \boldsymbol{k} \rangle . \tag{2.3.44}$$

Therefore, the right-hand side of (2.3.35) becomes

$$\frac{2}{3} \frac{1}{\beta} \sum_{\omega_n} \int \frac{\mathrm{d}^2 \boldsymbol{k}}{(2\pi)^2} \frac{g}{\omega_n^2 + \boldsymbol{k}^2 + m^2} . \tag{2.3.45}$$

The sum over ω_n can be performed by the standard technique of contour integration and, as a result, (2.3.45) becomes

$$\frac{3g}{4} \int \frac{\mathrm{d}^2 \boldsymbol{k}}{(2\pi)^2} \frac{1}{\omega(\boldsymbol{k})} \coth \left[\frac{\beta}{2} \omega(\boldsymbol{k}) \right] . \tag{2.3.46}$$

Here, the logarithmic divergence of the integral for $m \to 0$ at finite $\beta = 1/T$ signals that no long-range order at finite temperature emerges.

Therefore, the temperature dependence of $m(T)$ goes as follows. $m(T)$ is an increasing function that becomes zero for $T \to 0$ when $S > S_{\mathrm{c}}$, and reaches a finite value when $S < S_{\mathrm{c}}$. For $S > S_{\mathrm{c}}$, $m(T)$ is small for low temperature ($m(T) \ll T$ as can be seen in (2.3.48)). Approximating $\coth[\beta\omega(\boldsymbol{k})/2]$ by $2/\beta\omega(\boldsymbol{k})$, we obtain

$$\frac{3g}{2\beta} \int_{|\boldsymbol{k}|<\Lambda} \frac{\mathrm{d}^2 \boldsymbol{k}}{(2\pi)^2} \frac{1}{\boldsymbol{k}^2 + m(T)^2} \cong \frac{3g}{4\pi\beta} \ln \frac{\Lambda}{m(T)} . \tag{2.3.47}$$

For low temperature, we conclude therefore that

$$m(T) \cong \Lambda \, \mathrm{e}^{-4\pi/3gT} \cong \xi^{-1}(T) . \tag{2.3.48}$$

$m(T)$ shows an exponential behaviour with respect to the temperature. This strong temperature dependence of $\xi(T)$ for $T \to 0$ stands in striking contrast to the saturation for $S < S_{\mathrm{c}}$, and can be distinguished experimentally. For $S = 1/2$ it has been confirmed that $S > S_{\mathrm{c}}$.

In the three-dimensional case, quantum fluctuations are suppressed even more strongly than in the two-dimensional case. Therefore, long-range order exists. In the above discussion, we have seen that in low-dimensional quantum anti-ferromagnets, a rich variety of physical phenomena shows up owing to the interplay between quantum fluctuations and topological phases.

3. Strongly Correlated Electronic Systems

Up to the last chapter, the quantum spin system, especially the one-dimensional system, has been discussed. Problems related to this system are quite well understood. On the other hand, the topic of the present chapter – systems where charge and spin degrees of freedom are simultaneously present – is an active research field at present. New physical ideas such as the spin-charge separation are being exploited.

3.1 Models of Strongly Correlated Electronic Systems

Electron correlation is the phenomenon where electrons are aware of each other's motion owing to the repulsive Coulomb force. In this section, models describing the repulsive force between the electrons are introduced.

We start with the case where no interaction between the electrons is present. The single-electron state in the periodic potential $v(\boldsymbol{r})$ in the solid is described by Bloch waves. Writing $\phi_{n\boldsymbol{k}}(\boldsymbol{r})$ for this one-electron state, $\phi_{n\boldsymbol{k}}(\boldsymbol{r})$ obeys the Schrödinger equation

$$\left[-\frac{\hbar^2}{2m}\nabla^2 + v(\boldsymbol{r})\right]\phi_{n\boldsymbol{k}}(\boldsymbol{r}) = \varepsilon_n(\boldsymbol{k})\phi_{n\boldsymbol{k}}(\boldsymbol{r})\,.$$

We expand the electron field operator $\psi_\sigma^\dagger(\boldsymbol{r}), \psi_\sigma(\boldsymbol{r})$ in terms of the Bloch states

$$\begin{aligned}\psi_\sigma^\dagger(\boldsymbol{r}) &= \sum_{n,\boldsymbol{k}} \phi_{n\boldsymbol{k}}^*(\boldsymbol{r})C_{n\boldsymbol{k}\sigma}^\dagger\,,\\ \psi_\sigma(\boldsymbol{r}) &= \sum_{n,\boldsymbol{k}} \phi_{n\boldsymbol{k}}(\boldsymbol{r})C_{n\boldsymbol{k}\sigma}\,.\end{aligned} \tag{3.1.1}$$

Here, n specifies the band, and $\boldsymbol{k}$ is the wave number in the crystal in the first Brioullin zone. $C_{n\boldsymbol{k}\sigma}^\dagger(C_{n\boldsymbol{k}\sigma})$ are the creation (annihilation) operators of a $(n,\boldsymbol{k})$-Bloch electron state with spin σ. The anti-commutation relations are given by

$$\begin{aligned}\{C_{n\boldsymbol{k}\sigma}, C_{n'\boldsymbol{k}'\sigma'}\} &= \{C_{n\boldsymbol{k}\sigma}^\dagger, C_{n'\boldsymbol{k}'\sigma'}^\dagger\} = 0\\ \{C_{n\boldsymbol{k}\sigma}, C_{n'\boldsymbol{k}'\sigma'}^\dagger\} &= \delta_{nn'}\delta_{\boldsymbol{k}\boldsymbol{k}'}\delta_{\sigma\sigma'}\,.\end{aligned} \tag{3.1.2}$$

Now, we consider a cubic crystal with side length L and impose periodic boundary conditions at all sides. Then, we obtain with $\boldsymbol{m}$ being integer

$$\boldsymbol{k} = \frac{2\pi}{L}\boldsymbol{m}\,. \tag{3.1.3}$$

The possible number of $\boldsymbol{m}$ is the number N_0 of unit cells in the crystal. From the completeness relation of the Bloch states

$$\sum_{n,\boldsymbol{k}} \phi^*_{n,\boldsymbol{k}}(\boldsymbol{r})\phi_{n,\boldsymbol{k}}(\boldsymbol{r}') = \delta(\boldsymbol{r}-\boldsymbol{r}')\,, \tag{3.1.4}$$

we can deduce that

$$\{\psi_\sigma(\boldsymbol{r}), \psi^\dagger_{\sigma'}(\boldsymbol{r}')\} = \delta_{\sigma\sigma'}\delta(\boldsymbol{r}-\boldsymbol{r}')\,. \tag{3.1.5}$$

The Hamiltonian of the system can be expressed as

$$\begin{aligned} H_K &= \int \mathrm{d}\boldsymbol{r}\, \psi^\dagger_\sigma(\boldsymbol{r})\Big[-\frac{\hbar^2}{2m}\nabla^2 + v(\boldsymbol{r}) - \mu\Big]\psi_\sigma(\boldsymbol{r}) \\ &= \sum_{n,\boldsymbol{k},\sigma} (\varepsilon_n(\boldsymbol{k}) - \mu) C^\dagger_{n\boldsymbol{k}\sigma} C_{n\boldsymbol{k}\sigma} \end{aligned} \tag{3.1.6}$$

with μ being the chemical potential. The index K stands for kinetic energy.

In band theory, the metallic and insulator states can be distinguished by the relative position of the chemical potential μ, and the energy gap in the state density

$$D(\varepsilon) = \sum_{n,\boldsymbol{k}} \delta(\varepsilon - \varepsilon_n(\boldsymbol{k}))\,. \tag{3.1.7}$$

That is, $D(\varepsilon)$ consists of a large number of electron bands that are separated by energy gaps, as shown in Fig. 3.1 (the dispersion $\varepsilon_n(\boldsymbol{k})$ with index n corresponds to the specified band n in $D(\varepsilon)$ for the case when the energy gap is large enough. However, overlapping bands might also emerge where this is not the case).

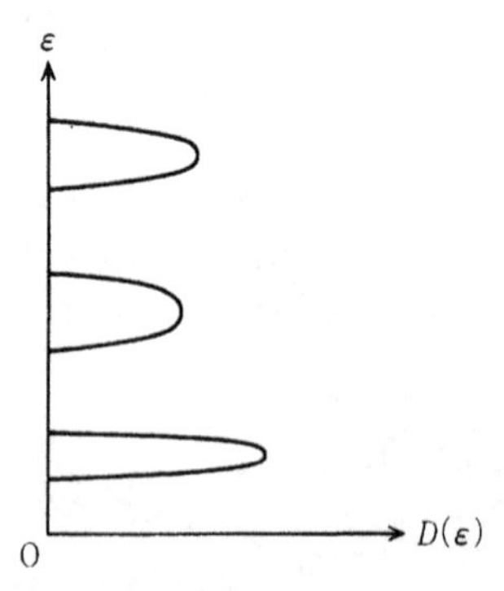

Fig. 3.1. Electron band structure in the solid

For the case when μ lies in a gap; that is, $D(\mu) = 0$, the system is an insulator, because no low-energy electron/hole pairs exist. For every n, there are N_0 $\boldsymbol{k}$-points in the first Brioullin zone, and with regard to the spin degeneracy, we conclude that $2N_0$ electrons can be packed into the first Brioullin zone. That is, bands that lie below μ are filled with an even number of $2N_0$ electrons, two in every elementary cell. On the other hand, $D(\mu) \neq 0$ characterizes the metallic state, and for $T = 0$, μ equals the so-called Fermi energy E_F. The surface defined by $\varepsilon_n(\boldsymbol{k}) = E_F$ in the $\boldsymbol{k}$-space is called the Fermi surface. The metallic properties at low temperature are determined by the electrons close to the Fermi surface.

Above, we introduced the central properties of so-called band theory. Although the interaction of the electrons has not been introduced explicitly, it is self-consistently incorporated in the mean field potential $v(\boldsymbol{r})$. In this sense, we define electron correlation as the effects of the repulsive force between electrons that cannot be incorporated in $v(\boldsymbol{r})$.

Now, the most fundamental Hamiltonian describing the interaction between the electrons is given by

$$H_{\mathrm{C}} = \frac{1}{2} \int \mathrm{d}\boldsymbol{r}\, \mathrm{d}\boldsymbol{r}'\, \psi_\sigma^\dagger(\boldsymbol{r}) \psi_{\sigma'}^\dagger(\boldsymbol{r}') \frac{e^2}{|\boldsymbol{r} - \boldsymbol{r}'|} \psi_{\sigma'}(\boldsymbol{r}') \psi_\sigma(\boldsymbol{r}) . \tag{3.1.8}$$

The full Hamiltonian is given by $H = H_K + H_{\mathrm{C}}$. The first issue arising is the long-range nature of the Coulomb force, causing infrared divergencies in a perturbative expansion in H_{C}. This problem can be solved in the framework of plasma oscillation and screening. That is, the Coulomb interaction causes collective excitation modes (plasma excitation), and after having separated this degree of freedom, the effective Coulomb force becomes a short-range force owing to the screening effect. Keeping this in mind, we will restrict the following discussion to a short-range force. However, attention has to be paid to the fact that screening occurs only in the metallic phase, and therefore it is principally dangerous to discuss the phase transition between the metallic and the insulator phase in a model based on a short-range force. Indeed, in the original work of Mott, just due to the long-range characteristic, he showed that the phase transition is first order.

To build a model for the short-range repulsive force, it is better to switch from the picture of Bloch waves $\phi_{n\boldsymbol{k}}(\boldsymbol{r})$, being widely spread in the real space, to Wannier orbits $\phi_n(\boldsymbol{r} - \boldsymbol{R}_l)$ that are localized around the position $\boldsymbol{R}_l$ of each atom.

$$\phi_n(\boldsymbol{r} - \boldsymbol{R}_l) = \frac{1}{\sqrt{L^d}} \sum_{\boldsymbol{k}} \mathrm{e}^{-\mathrm{i}\boldsymbol{k} \cdot \boldsymbol{R}_l} \phi_{n\boldsymbol{k}}(\boldsymbol{r}) . \tag{3.1.9}$$

The expansion of the operator of the electron field now reads

$$\psi_\sigma^\dagger(\boldsymbol{r}) = \sum_{n,l} \phi_n^*(\boldsymbol{r} - \boldsymbol{R}_l) C_{nl\sigma}^\dagger ,$$
$$\psi_\sigma(\boldsymbol{r}) = \sum_{n,l} \phi_n(\boldsymbol{r} - \boldsymbol{R}_l) C_{nl} . \tag{3.1.10}$$

In this framework, H_K (3.1.6) becomes

$$H_K = - \sum_{n,l,l',\sigma} t_{l-l'}^{(n)} C_{nl\sigma}^\dagger C_{nl'\sigma} - \mu \sum_{n,l,\sigma} C_{nl\sigma}^\dagger C_{nl\sigma} . \tag{3.1.11}$$

Here,

$$t_{l-l'}^{(n)} = \frac{-1}{L^d} \sum_{\boldsymbol{k}} \exp[-\mathrm{i}\boldsymbol{k} \cdot (\boldsymbol{R}_l - \boldsymbol{R}_{l'})] \varepsilon_n(\boldsymbol{k}) . \tag{3.1.12}$$

is the so-called transfer integral from site l to l'.

For the case when only one band contributes to the conduction, we omit the band index and obtain the single-band, tight-binding model

$$H_K = - \sum_{l,l',\sigma} t_{l-l'} C_{l\sigma}^\dagger C_{l'\sigma} - \mu \sum_{l} C_{l\sigma}^\dagger C_{l\sigma} . \tag{3.1.13}$$

The short-range repulsive force is expressed by

$$H_U = U \sum_{l} n_{l\uparrow} n_{l\downarrow} \tag{3.1.14}$$

with $n_{l\sigma} = C_{l\sigma}^\dagger C_{l\sigma}$ being the electron number with spin σ. U is the repulsive energy between the electrons. The sum $H = H_K + H_U$ is the so-called Hubbard model. This model is the most fundamental model when considering electron correlation. It is also applied to the description of d-electrons in transition metals oxide.

We now give a qualitative picture of the physics that is described by H_K and H_U. H_K represents the wave nature of the electron to spread over the whole crystal; in other words, the mobility. If we dealt with bosons, H_K would cause Bose condensation, and create a coherent state where the quantum mechanical phase is macroscopically fixed. However, because electrons are fermions, owing to the negative sign occurring when two particles are interchanged, a kind of frustration occurs, and it is impossible that all electrons occupy the lowest energy state. This leads to Fermi degeneracy.

On the other hand, H_U fixes the electron number $n_{l\sigma}$, and describes an interaction that stresses the nature of the particle. That is, H_U tends to localize and insulate the electrons. The competition between H_U and H_K is the basic physical picture of the Hubbard model.

However, another important issue is contained in the Hubbard model. In the above description, we focused on the charge degree of freedom of the electron. However, the electron has one more degree of freedom, being the spin. The spin dominates the magnetic properties (magnetism) of the system. From this point of view, H_K tries to undo the effect of the spin by Fermi degeneracy. That is, except for the region close to the Fermi surface, every spin of the electron is accompanied by the inverse, and the system is magnetically almost inert. This fact can easily be understood by considering the susceptibility $\chi(T)$. The susceptibility of free spins obeys the Curie law

$$\chi_{\text{Curie}}(T) = \frac{S(S+1)\mu_{\text{B}}^2}{3T}. \quad (3.1.15)$$

μ_{B} is the Bohr magneton, and S is the spin (for electrons $S = 1/2$). On the other hand, the susceptibility of metals following from the Hamiltonian H_K is given by the Pauli susceptibility

$$\chi_{\text{Pauli}} = 2\mu_{\text{B}}^2 D(E_{\text{F}}), \quad (3.1.16)$$

which is independent of temperature for low temperatures.

This can be understood as follows. At temperature T, only the fraction $\simeq T/E_{\text{F}}$ of the electrons around the Fermi surface can contribute to the magnetism. When only these electrons obey the Curie law χ_{Curie} for free spins, the result is

$$\chi(T) \sim \chi_{\text{Curie}}(T) \times \frac{T}{E_{\text{F}}} \sim \frac{\mu_{\text{B}}^2}{E_{\text{F}}}, \quad (3.1.17)$$

a temperature-independent susceptibility. With regard to $D(E_{\text{F}}) \propto E_{\text{F}}^{-1}$, we conclude that (3.1.17) is consistent with (3.1.16). In real space, this can be interpreted as follows: Because the spin $\uparrow$ and spin $\downarrow$ move independently of one another, the spin cancels out.

On the other hand, H_U expresses the energy cost when at the same site, spin $\uparrow$ and spin $\downarrow$ are present. Therefore, the cancellation mentioned above is reduced and spin moments are induced under the influence of H_U. This can be seen more easily when we rewrite H_U in the following manner:

$$H_U = U\sum_l n_{l\uparrow}n_{l\downarrow} = \frac{U}{2}\sum_l (n_{l\uparrow} + n_{l\downarrow}) - \frac{U}{2}\sum_l (n_{l\uparrow} - n_{l\downarrow})^2. \quad (3.1.18)$$

Here, we used $n_{l\sigma}^2 = n_{l\sigma}$. Because $(n_{l\uparrow} - n_{l\downarrow})$ is twice the z-component of the spin, the minus sign in the second term of H_U signifies that H_U induces the z-component. However, because it is not necessary to choose a specific spin direction, using the spin operator of the electron

$$\boldsymbol{S}_l = \frac{1}{2}\sum_{\alpha,\beta} C_{l\alpha}^\dagger \boldsymbol{\sigma}_{\alpha\beta} C_{l\beta}, \quad (3.1.19)$$

(3.1.18) can be expressed as

$$H_U = \frac{U}{2}\sum_l (n_{l\uparrow} + n_{l\downarrow}) - \frac{2U}{3}\sum_l \boldsymbol{S}_l^2, \quad (3.1.20)$$

which is an isotropic expression in the spin space. In such a way, the repulsive force between the electrons leads to a spin moment, finally causing magnetism. The basic physics of strongly correlated electronic systems is the competition between the two tendencies of the electron to spread out as a wave and to localize as a particle, combined with magnetism. That is, the interplay of the spin and the charge degree of freedom is the central issue.

Above, we discussed the case of a single band, but there are also cases where many bands become important. For example, for rare earth compounds, where at the same time conduction electrons and strongly localized f electrons are present, the following periodic Anderson model has been studied:

$$
\begin{aligned}
H = &\sum_{k,\sigma}(\varepsilon_k - \mu)C^\dagger_{k\sigma}C_{k\sigma} + E_\mathrm{f}\sum_{k,\sigma} f^\dagger_{k\sigma}f_{k\sigma} + U\sum_{l} n_{\mathrm{f}l\uparrow} - n_{\mathrm{f}l\downarrow} \\
&-\sum_{k,\sigma} V_k\left(C^\dagger_{k,\sigma}f_{k\sigma} + f^\dagger_{k\sigma}C_{k\sigma}\right). \qquad (3.1.21)
\end{aligned}
$$

Here, the repulsive force U acts only on the f electrons, which are hybridized with the itinerant conduction electrons through V_k.

Intensive studies have been done on the dilute magnetic impurity ions in metals as a problem of local electron correlation. The Kondo problem is the central issue in this problem. The basic model is the Anderson Hamiltonian for one contamination particle

$$
\begin{aligned}
H = &\sum_{k,\sigma}(\varepsilon_k - \mu)C^\dagger_{k\sigma}C_{k\sigma} + E_\mathrm{f}\sum_{\sigma} f^\dagger_{\sigma}f_{\sigma} + Un_{\mathrm{f}\uparrow}n_{\mathrm{f}\downarrow} \\
&-\frac{1}{\sqrt{N_0}}\sum_{k,\sigma} V_k\left(C^\dagger_{k\sigma}f_{\sigma} + f^\dagger_{\sigma}C_{k\sigma}\right). \qquad (3.1.22)
\end{aligned}
$$

Here, N_0 is the number of lattice points in the crystal.

The above models have been defined for arbitrary U, and it is possible to derive effective Hamiltonians in the limit of strong repulsion $U \to \infty$. First, we consider the Hubbard model. An important parameter in the tight-binding model is the filling factor ν of the electrons. ν is the ratio between the number of electrons N_e and the number of lattice points N_0, $\nu = N_\mathrm{e}/N_0$. $\nu = 0$ and $\nu = 2$ correspond to the totally empty and the totally filled band, and in both cases the band is non-conducting. For other values of ν, metallic properties should emerge, whereas the value $\nu = 1$ plays a special role. $\nu = 1$ just corresponds to the case where half of the band is filled, and around one lattice point one electron is present (so-called half-filling). In this case, for U large enough in the Hubbard model, the system becomes an insulator (Mott insulator). We will explain this fact using Fig. 3.2.

In the limit of large U, H_K can be ignored as the zeroth order approximation. Then, as shown in Fig. 3.2, groups of states with energy gap U arise. This follows from the fact that H_U is given by U times the number of sites

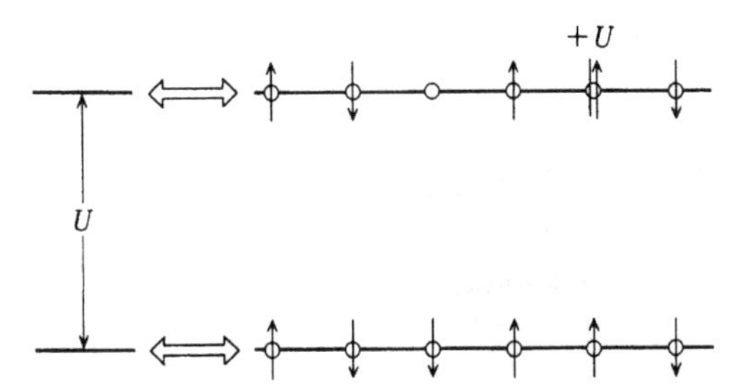

Fig. 3.2. The energy spectrum of the Mott insulator. Ignoring the electron hopping t, the energy is determined by the number of double-occupied sites only. In the lowest energy state, one electron sits at every site. Owing to the spin degree of freedom, this state is 2^{N_0} times degenerate

occupied by two electrons. In particular, the lowest lying state has no site occupied by two electrons, but at every site there is just one electron. Because every site has two states, owing to two possible spin positions, this lowest level is 2^{N_0}-fold degenerated. The projection operator on this group of states is given by $P = \Pi_l(1 - n_{l\uparrow}n_{l\downarrow})$, and we call the corresponding Hilbert space $\mathcal{H}$.

We want to derive the effective Hamiltonian describing the states of $\mathcal{H}$. We assume that $|t_{l-l'}| \ll U$ and derive this Hamiltonian using a perturbative expansion in $t_{l-l'}$. To do so, we start with the Schrödinger equation

$$H\Psi = E\Psi\,. \tag{3.1.23}$$

Using P and $Q = 1 - P$, we can write

$$H(P+Q)\Psi = E(P+Q)\Psi\,. \tag{3.1.24}$$

Notice that $P^2 = P$, $Q^2 = Q$, and $PQ = QP = 0$. The state $P\Psi$ is an element of $\mathcal{H}$, and $Q\Psi$ contains only states lying energetically about U or more higher. Therefore, we want to eliminate $Q\Psi$ in (3.1.24) and try to derive an eigenvalue problem for $P\Psi$. Multiplying (3.1.24) from the left with Q, after some transformations we obtain

$$(QHQ - E)Q\Psi = -QHP\Psi\,. \tag{3.1.25}$$

Inserting $Q\Psi = -(QHQ - E)^{-1}QHP\Psi$ in (3.1.24), we obtain

$$(H - H(QHQ - E)^{-1}QH)P\Psi = E(P+Q)\Psi\,. \tag{3.1.26}$$

Multiplying from the left with P, the result is

$$[PHP - PHQ(QHQ - E)^{-1}QHP]P\Psi = EP\Psi\,. \tag{3.1.27}$$

Up to this point, we have considered the general case, which will now be applied to the Hubbard model $H = H_K + H_U$ with $P = \Pi_l(1 - n_{l\uparrow}n_{l\downarrow})$

$$PHP =: PH_U P = 0 \tag{3.1.28}$$

$$PHQ = PH_K Q\,, \qquad QHP = QH_K P\,. \tag{3.1.29}$$

Considering the matrix elements PH_KQ and QH_KP, a state of $\mathcal{H}$ which is multiplied once with H_K belongs to the group of states where only one site is occupied by two electrons. Now, at the left-hand side of (3.1.27), QHQ in $(QHQ-E)^{-1}$ is of zeroth order in $t_{l-l'}$ and can be replaced by U. This can be done because PHQ and QHP are already of second order with respect to $t_{l-l'}$. In the same way, E can be replaced by the lowest order in $t_{l-l'}$; that is $E = 0$. We conclude that the effective Hamiltonian H_{eff} in the Hilbert space $\mathcal{H}$ is given by

$$H_{\text{eff}} = -\frac{(PH_KQ)(QH_KP)}{U} = -P\frac{H_K^2}{U}P\,. \tag{3.1.30}$$

The explicit calculation leads to

$$PH_K^2P = \sum_{\substack{l_1,l_1'\\ l_2,l_2'}} \sum_{\sigma_1,\sigma_2} t_{l_1-l_1'} t_{l_2-l_2'} P C_{l_1\sigma_1}^\dagger C_{l_1'\sigma_1} C_{l_2\sigma_2}^\dagger C_{l_2'\sigma_2} P\,. \tag{3.1.31}$$

Because P is multiplied on both sides, an electron that is annihilated at l_2' must in every case reappear at the same place. With regard to $t_{l_1-l_1'=0} = 0$, only $l_1 = l_2'$ remains. In the same way, only $l_1' = l_2$ remains, and (3.1.31) can be written as

$$\begin{aligned} PH_K^2P &= \sum_{\substack{l_1,l_2\\ \sigma_1,\sigma_2}} t_{l_1-l_2} t_{l_2-l_1} P C_{l_1\sigma_1}^\dagger C_{l_2\sigma_1} C_{l_2\sigma_2}^\dagger C_{l_1\sigma_2} P \\ &= \sum_{\substack{l_1,l_2\\ \sigma_1,\sigma_2}} |t_{l_1-l_2}|^2 P C_{l_1\sigma_1}^\dagger C_{l_1\sigma_2} P \cdot P C_{l_2\sigma_1} C_{l_2\sigma_2}^\dagger P\,. \end{aligned} \tag{3.1.32}$$

Now, we use the identity

$$\begin{aligned} C_{l\sigma_1}^\dagger C_{l\sigma_2} &= \frac{1}{2}\delta_{\sigma_1\sigma_2}(n_{l\uparrow} + n_{l\downarrow}) + \boldsymbol{S}_l \cdot \boldsymbol{\sigma}_{\sigma_2\sigma_1}\,, \\ C_{l\sigma_1} C_{l\sigma_2}^\dagger &= \delta_{\sigma_1\sigma_2}\left(1 - \frac{n_{l\uparrow} + n_{l\downarrow}}{2}\right) - \boldsymbol{S}_l \cdot \boldsymbol{\sigma}_{\sigma_1\sigma_2}\,. \end{aligned} \tag{3.1.33}$$

Here, $\boldsymbol{S}_l$ is the spin of the electron as defined in (3.1.19). Inserting this identity into (3.1.32), we obtain

$$\begin{aligned} PH_K^2P &= \sum_{\substack{l_1,l_2\\ \sigma_1,\sigma_2}} |t_{l_1-l_2}|^2 \left(\frac{\delta_{\sigma_1\sigma_2}}{2} + \boldsymbol{S}_{l_1} \cdot \boldsymbol{\sigma}_{\sigma_2\sigma_1}\right)\left(\frac{\delta_{\sigma_2\sigma_1}}{2} - \boldsymbol{S}_{l_2} \cdot \boldsymbol{\sigma}_{\sigma_1\sigma_2}\right) \\ &= \sum_{l_1,l_2} |t_{l_1-l_2}|^2 \operatorname{Tr}\left[\left(\frac{1}{2} + \sum_\alpha S_{l_1}^\alpha \sigma^\alpha\right)\left(\frac{1}{2} - \sum_\beta S_{l_2}^\beta \sigma^\beta\right)\right] \\ &= \sum_{l_1,l_2} |t_{l_1-l_2}|^2 \left(\frac{1}{2} - 2\boldsymbol{S}_{l_1} \cdot \boldsymbol{S}_{l_2}\right), \end{aligned} \tag{3.1.34}$$

where we used $\mathrm{Tr}[\sigma^\alpha \sigma^\beta] = 2\delta^{\alpha\beta}$.

$$H_{\text{eff}} = \frac{2}{U} \sum_{l_1,l_2} |t_{l_1-l_2}|^2 \left(\boldsymbol{S}_{l_1} \cdot \boldsymbol{S}_{l_2} - \frac{1}{4} \right) . \tag{3.1.35}$$

With

$$J_{l_1,l_2} = \frac{2}{U} |t_{l_1-l_2}|^2 > 0 , \tag{3.1.36}$$

we obtain finally

$$H_J = \sum_{l_1,l_2} J_{l_1,l_2} \left(\boldsymbol{S}_{l_1} \cdot \boldsymbol{S}_{l_2} - \frac{1}{4} \right) , \tag{3.1.37}$$

which is nothing more than the anti-ferromagnetic Heisenberg model.

Here, the reason why anti-ferromagnetic interaction arises becomes evident. The calculation performed beginning from (3.1.23) is a perturbative calculation up to second order of the energy gain, as demonstrated in Fig. 3.3. In the case when spin l_1 and spin l_2 are parallel, the intermediate state, where two electrons with parallel spin are at the same site, would be forbidden owing to the Pauli principle. Therefore, the spins align anti-parallel.

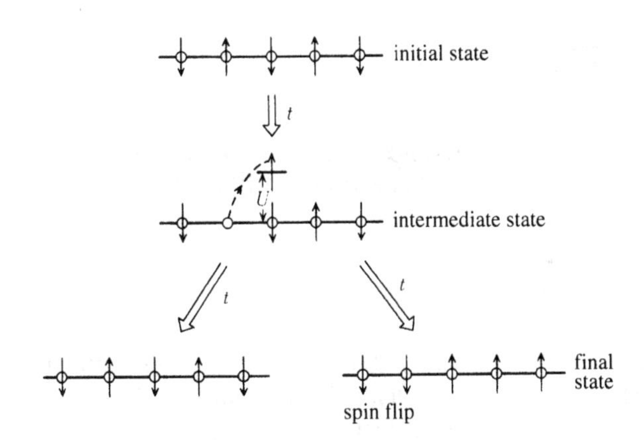

Fig. 3.3. A process caused by the exchange interaction in second-order perturbation theory

Above, we defined the effective Hamiltonian for a restricted part $\mathcal{H}$ of the Hilbert space, which leads to a constraint. In the present case, the constraint is that one electron is present at every site

$$\sum_{\sigma} C_{l\sigma}^\dagger C_{l\sigma} = 1 . \tag{3.1.38}$$

Under this constraint, the spin operator (3.1.19) $\boldsymbol{S}_l$ satisfies the spin commutation relation (1.1.2). In various situations, constraints will appear in the theory of strongly correlated electronic systems. The above constraint is the most typical example.

Because in the above discussion the states in $\mathcal{H}$ have been considered, the system is always an insulator. This is evident because when charge flows, the energy gap U must be surpassed. In such a way, the insulator realized owing to $\nu = 1$ is purely caused by electron correlation, and is the so-called Mott insulator. Its low-energy excitations are due to the spin degree of freedom, as described by (3.1.37). However, when ν differs slightly from one, low-energy excitations due to the charge will also occur. For example, when the number of electrons is shifted from $\nu = 1$ by x (doping with holes with rate x), then the system has $\nu = 1 - x$ electrons. Then, the sites with occupation number zero are allowed in the Hilbert space $\mathcal{H}$.

The effective Hamiltonian of this system differs in three ways from the Hamiltonian derived in (3.1.37).

(i) PH_KP is different from zero

$$PH_KP = -\sum_{\substack{l,l' \\ \sigma}} t_{l-l'} PC_{l\sigma}^{\dagger}P \cdot PC_{l'\sigma}P \,. \tag{3.1.39}$$

This describes the motion of the holes.

(ii) The operator $n_l = n_{l\uparrow} + n_{l\downarrow}$ in (3.1.33) is not fixed to be one, and therefore the interchange interaction term becomes (ignoring coefficients)

$$H_J = \sum_{l_1,l_2} J_{l_1,l_2}\left(\boldsymbol{S}_{l_1} \cdot \boldsymbol{S}_{l_2} + \frac{1}{4} n_{l_1} n_{l_2}\right). \tag{3.1.40}$$

(iii) The constraint excludes only the presence of double occupancy, but allows the presence of holes. Therefore, instead of (3.1.38), we obtain the inequality

$$\sum_{\sigma} C_{l\sigma}^{\dagger} C_{l\sigma} \leqq 1 \,. \tag{3.1.41}$$

The sum of the Hamiltonian (3.1.39) and (3.1.40) $H = PH_KP + H_J$ under the constraint (3.1.41) is the so-called t-J model, being the most fundamental model describing high-temperature superconductors as well as other doped Mott insulators.

3.2 Spin-Charge Separation in One Dimension

As has been mentioned in the previous section, the central theme in strongly correlated electronic systems is the interplay between the degree of freedom of spin and charge. In the one-dimensional electron system, the striking feature of spin–charge separation occurs, which is the characteristic feature of non-Fermi liquids. This will be discussed in what follows.

In Sect. 2.1, spinless fermions have been introduced. The idea we started with was the 'Fermi wave number expressed as an operator'. That is, when $k_F + \delta k_F^R(x,\tau)$ and $-k_F - \delta k_F^L(x,\tau)$ are the Fermi wave numbers of the right-moving and left-moving fermions, the electron charge density is given by $\rho^R(x,\tau) = 1/(2\pi)\delta k_F^R(x,\tau)$ and $\rho^L(x,\tau) = 1/(2\pi)\delta k_F^L(x,\tau)$. That is, the variation of the Fermi wave number is nothing more than the charge density fluctuation. Because the phase of the electron wave is given by the x integration of the Fermi wave number, we can write

$$\begin{aligned} \psi_{R\sigma}(x) &\propto \exp\left[ik_F x + \mathrm{i}\int_{-\infty}^{x} \delta k_F^{R\sigma}(x',\tau)\,\mathrm{d}x'\right], \\ \psi_{L\sigma}(x) &\propto \exp\left[-\mathrm{i}k_F x - \mathrm{i}\int_{-\infty}^{x} \delta k_F^{L\sigma}(x',\tau)\,\mathrm{d}x'\right], \end{aligned} \tag{3.2.1}$$

where we introduced the spin label σ.

Here, four different fields $\delta k_F^{R\sigma}(x,\tau)$ and $\delta k_F^{L\sigma}(x,\tau)$ are on stage and, correspondingly, four different types of densities are defined:

$$\begin{aligned}
&\text{(i) electron density:} & \rho(x) &= \frac{1}{2\pi}\sum_\sigma \left(\delta k_F^{R\sigma}(x) + \delta k_F^{L\sigma}(x)\right) \\
&& &= \sum_\sigma \left(\rho_\sigma^R(x) + \rho_\sigma^L(x)\right), \\
&\text{(ii) electron current density:} & j(x) &= \frac{1}{2\pi}\sum_\sigma \left(\delta k_F^{R\sigma}(x) - \delta k_F^{L\sigma}(x)\right) \\
&& &= \sum_\sigma \left(\rho_\sigma^R(x) - \rho_\sigma^l(x)\right), \\
&\text{(iii) spin density:} & 2s(x) &= \frac{1}{2\pi}\sum_\sigma \sigma\left(\delta k_F^{R\sigma}(x) + \delta k_F^{L\sigma}(x)\right) \\
&& &= \sum_\sigma \sigma\left(\rho_\sigma^R(x) + \rho_\sigma^L(x)\right), \\
&\text{(iv) spin current density:} & 2j_s(x) &= \frac{1}{2\pi}\sum_\sigma \sigma\left(\delta k_F^{R\sigma}(x) - \delta k_F^{L\sigma}(x)\right) \\
&& &= \sum_\sigma \sigma\left(\rho_\sigma^R(x) - \rho_\sigma^L(x)\right).
\end{aligned} \tag{3.2.2}$$

These densities are the slowly varying part in space. They must be distinguished from the $2k_F$ component that will be introduced in what follows. The commutation relations are given by

$$\begin{aligned} \left[\rho_\sigma^R(-p), \rho_{\sigma'}^R(p')\right] &= -\frac{p}{2\pi}\delta_{pp'}\delta_{\sigma\sigma'}, \\ \left[\rho_\sigma^L(-p), \rho_{\sigma'}^L(p')\right] &= \frac{p}{2\pi}\delta_{pp'}\delta_{\sigma\sigma'}, \end{aligned}$$

where as in (2.1.19), $p > 0$. We conclude that

$$
\begin{aligned}
[\rho(-p), \jmath(p')] &= -\frac{2}{\pi} p \delta_{pp'} , \\
[2s(-p), 2\jmath_s(p')] &= -\frac{2}{\pi} p \delta_{pp'} .
\end{aligned}
\tag{3.2.3}
$$

All other combinations commute. We see that $\rho(x)$ and $\jmath(x)$ as well as $s(x)$ and $\jmath_s(x)$ are canonical conjugates, respectively.

As we have learnt in the discussion of Sect. 2.1, we express the Hamiltonian in terms of these fields. We consider the one-dimensional Hubbard model

$$
H = H_K + H_U = -t \sum_{l,\sigma} (C^\dagger_{l,\sigma} C_{l+1,\sigma} + C^\dagger_{l+1,\sigma} C_l) + U \sum_l n_{l\uparrow} n_{l\downarrow} . \tag{3.2.4}
$$

Using a variation of the calculation (2.1.50)–(2.1.53), we can express H_K as

$$
H_K^{\mathrm{B}} = 4\pi t \sum_{\substack{p>0 \\ \sigma}} \left[\rho^\sigma_{\mathrm{R}}(-p) \rho^\sigma_{\mathrm{R}}(p) + \rho^\sigma_{\mathrm{L}}(-p) \rho^\sigma_{\mathrm{L}}(p) \right] . \tag{3.2.5}
$$

This equals

$$
H_K^{\mathrm{B}} = \frac{\pi t}{2} \int \mathrm{d}x \left[\rho(x)^2 + \jmath(x)^2 + (2s(x))^2 + (2\jmath_s(x))^2 \right] . \tag{3.2.6}
$$

Next, we consider H_U. This term can be expressed in many different ways.

$$
\begin{aligned}
H_U = U \sum_l n_{l\uparrow} n_{l\downarrow} &= \frac{U}{2} \sum_l \left[(n_{l\uparrow} + n_{l\downarrow})^2 - (n_{l\uparrow} + n_{l\downarrow}) \right] \\
&= \frac{U}{2} \sum_l \left[(n_{l\uparrow} + n_{l\downarrow}) - (n_{l\uparrow} - n_{l\downarrow})^2 \right] \\
&= \frac{U}{4} \sum_l \left[(n_{l\uparrow} + n_{l\downarrow})^2 - (n_{l\uparrow} - n_{l\downarrow})^2 \right] .
\end{aligned}
\tag{3.2.7}
$$

Owing to $n^2_{l\sigma} = n_{l\sigma}$, all the above equations are exact. However, the description of the physics changes. Because $n_{l\uparrow} + n_{l\downarrow} = \rho_l$ is the charge, and $n_{l\uparrow} - n_{l\downarrow} = 2s_l$ the spin, there are in the continuum limit the options to write $\rho(x)$ as well as $s(x)$.

Here, we use the third line where charge and spin occur in a symmetric way. The continuum limit of H_U is then given by

$$
H_U^{\mathrm{B}} = \frac{U}{4} \int \mathrm{d}x \left[\rho(x)^2 - (2s(x))^2 \right] , \tag{3.2.8}
$$

where the part that can be absorbed in the chemical potential has been omitted. This term corresponds to forward scattering. The sum H_0 of H_K^{B} and H_U^{B} can be written using

$$
H_{\mathrm{charge}} = \int \mathrm{d}x \left[\left(\frac{\pi t}{2} + \frac{U}{4} \right) \rho(x)^2 + \frac{\pi l}{2} \jmath(x)^2 \right] \tag{3.2.9}
$$

and

$$H_{\text{spin}} = \int \mathrm{d}x \left[\left(\frac{\pi t}{2} - \frac{U}{4} \right) (2s(x))^2 + \frac{\pi t}{2} (2j_\text{s}(x))^2 \right], \tag{3.2.10}$$

with $H_0 = H_{\text{charge}} + H_{\text{spin}}$.

In such a manner, the Hamiltonian can be expressed as a sum, and because also the commutation relations decouple like (3.2.3), the charge fluctuations and the spin fluctuations are totally independent from one another. This is the so-called spin-charge separation in one dimension.

To express the Hamiltonians in a clearer manner, we introduce the following phase fields

$$\begin{aligned} \rho(x) &= \frac{1}{\pi} \partial_x \phi_\text{c}(x) \,, \\ j(x) &= \frac{1}{\pi} \partial_x \theta_\text{c}(x) \,, \\ 2s(x) &= \frac{1}{\pi} \partial_x \phi_\text{s}(x) \,, \\ 2j_\text{s}(x) &= \frac{1}{\pi} \partial_x \theta_\text{s}(x) \,. \end{aligned} \tag{3.2.11}$$

The commutation relations are due to (3.2.3)

$$\begin{aligned} [\phi_\text{c}(x), \theta_\text{c}(x')] &= \mathrm{i}\pi \, \mathrm{sgn}(x - x') \,, \\ [\phi_\text{s}(x), \theta_\text{s}(x')] &= \mathrm{i}\pi \, \mathrm{sgn}(x - x') \,. \end{aligned} \tag{3.2.12}$$

Then, H_{charge} and H_{spin} can be expressed as

$$H_{\text{charge}} = \frac{v_\rho}{4\pi} \int \mathrm{d}x \left[\frac{1}{K_\rho} (\partial_x \phi_\text{c})^2 + K_\rho (\partial_x \theta_\text{c})^2 \right], \tag{3.2.13}$$

$$H_{\text{spin}} = \frac{v_\sigma}{4\pi} \int \mathrm{d}x \left[\frac{1}{K_\sigma} (\partial_x \phi_\text{s})^2 + K_\sigma (\partial_x \theta_\text{s})^2 \right] \tag{3.2.14}$$

with

$$v_\rho = 2\sqrt{t\left(t + \frac{U}{2\pi}\right)}, \qquad K_\rho = \sqrt{\frac{t}{t + U/2\pi}}, \tag{3.2.15}$$

$$v_\sigma = 2\sqrt{t\left(t - \frac{U}{2\pi}\right)}, \qquad K_\sigma = \sqrt{\frac{t}{t - U/2\pi}}. \tag{3.2.16}$$

Here, v_ρ and v_σ are the velocities $\omega = v_\rho k$ and $\omega = v_\sigma k$ of the dispersion relations of ϕ_c and ϕ_s, respectively. Owing to the effect of U, v_ρ and v_σ differ from one another. This is the 'visible effect' of spin-charge separation.

K_ρ and K_σ are the coefficients determining the strength of the rivals charge and current. For $K_{\sigma,\rho} > 1$ the current and for $K_{\sigma,\rho} < 1$ the charge density, respectively, win and tend to be fixed.

Following (3.2.1), the electron operator is given by

$$\psi_{\mathrm{R}\sigma} \propto \exp\left[\mathrm{i}k_{\mathrm{F}}x + \frac{\mathrm{i}}{2}[\phi_{\mathrm{c}} + \theta_{\mathrm{c}} + \sigma(\phi_{\mathrm{s}} + \theta_{\mathrm{s}})]\right],$$
$$\psi_{\mathrm{L}\sigma} \propto \exp\left[-\mathrm{i}k_{\mathrm{F}}x - \frac{\mathrm{i}}{2}[\phi_{\mathrm{c}} - \theta_{\mathrm{c}} + \sigma(\phi_{\mathrm{s}} - \theta_{\mathrm{s}})]\right]. \tag{3.2.17}$$

Up to now, we have discussed the action up to second order of the bosonic fields. However, similar to Sect. 2.1, the non-linear trigonometric term plays an important role. One example is the term describing the so-called Umklapp scattering

$$H_{\mathrm{Umklapp}} = 2U \int \mathrm{d}x \{\mathrm{e}^{\mathrm{i}Gx}\psi^\dagger_{\mathrm{R}\uparrow}(x)\psi^\dagger_{\mathrm{R}\downarrow}(x)\psi_{\mathrm{L}\downarrow}(x)\psi_{\mathrm{L}\uparrow}(x) + \mathrm{h.c.}\}, \tag{3.2.18}$$

where G is the vector of the reciprocal lattice. Inserting (3.2.17), with $g_3 \propto U$

$$H_{\mathrm{Umklapp}} = g_3 \int \mathrm{d}x \cos[2\phi_{\mathrm{c}}(x) + (4k_{\mathrm{F}} - G)x], \tag{3.2.19}$$

the Umklapp scattering can be expressed in terms of $\phi_{\mathrm{c}}(x)$ only.

Because $\phi_{\mathrm{c}}(x)$ is a slowly varying field, in the long-range, low-energy limit, this field is only relevant for $4k_{\mathrm{F}} = G$; that is, in the half-filled case. Otherwise, the term $(4k_{\mathrm{F}} - G)x$ would cause sign changes in the cos and therefore cancellation. Using the analysis of Sect. 2.1, we can show that this term is marginal when K_ρ in (3.2.13) equals one, relevant for $K_\rho < 1$ and irrelevant for $K_\rho > 1$. That is, for $U > 0$, g_3 scales to larger and larger values, and for $U < 0$, the scaling tends to zero. As a result, in the repulsive Hubbard model, the degree of freedom ϕ_{c} of the charge is fixed due to the cos potential, and therefore a gap in the excitation spectrum arises. This is the description of the one-dimensional Mott insulator. On the other hand, for attractive interaction, the spectrum of the degree of freedom of the charge is gapless.

On the other hand, concerning the spin degree of freedom, the following term expresses the non-linear interaction, called backwards scattering:

$$H_{\mathrm{b}} = U \int \mathrm{d}x \{\psi^\dagger_{\mathrm{L}\sigma}(x)\psi^\dagger_{\mathrm{R}\sigma'}(x)\psi_{\mathrm{L}\sigma'}(x)\psi_{\mathrm{R}\sigma}(x) + \mathrm{h.c.}\}$$
$$= g_1 \int \mathrm{d}x \cos 2\phi_{\mathrm{s}}, \tag{3.2.20}$$

with $g_1 \propto U$. K_σ and g_1 should obey the scaling law derived in Sect. 2.1, and at the fixed point, $SU(2)$ symmetry arises, which corresponds to $K_\sigma = 1$. This can be seen as follows. When we calculate the $2k_{\mathrm{F}}$-component spin density using (3.2.17), we obtain

$$S^z(x)|_{2k_{\mathrm{F}}} \sim \psi^\dagger_{\mathrm{R}\uparrow}(x)\psi_{\mathrm{L}\uparrow}(x) - \psi^\dagger_{\mathrm{R}\downarrow}(x)\psi_{\mathrm{L}\downarrow}(x) \propto \mathrm{e}^{-2\mathrm{i}k_{\mathrm{F}}x}\,\mathrm{e}^{-\mathrm{i}\phi_{\mathrm{c}}}\sum_{\sigma}\sigma\,\mathrm{e}^{-\mathrm{i}\sigma\phi_{\mathrm{s}}}\,, \tag{3.2.21}$$

$$S^+(x)|_{2k_{\mathrm{F}}} \sim \psi^\dagger_{\mathrm{R}\uparrow}(x)\psi_{\mathrm{L}\downarrow}(x) \propto \mathrm{e}^{-2\mathrm{i}k_{\mathrm{F}}x}\,\mathrm{e}^{-\mathrm{i}\phi_{\mathrm{c}}}\,\mathrm{e}^{-\mathrm{i}\theta_{\mathrm{s}}}\,, \tag{3.2.22}$$

and in the same manner as in Sect. 2.1, when calculating the correlation function, its asymptotic form is given by

$$\langle S^z(x,\tau)S^z(0,0)\rangle_{2k_{\mathrm{F}}} \sim \frac{1}{(x^2+v_\rho^2\tau^2)^{K_\rho/2}}\frac{1}{(x^2+v_\sigma^2\tau^2)^{K_\sigma/2}}\,, \tag{3.2.23}$$

$$\langle S^+(x,\tau)S^-(0,0)\rangle_{2k_{\mathrm{F}}} \sim \frac{1}{(x^2+v_\rho^2\tau^2)^{K_\rho/2}}\frac{1}{(x^2+v_\sigma^2\tau^2)^{1/(2K_\sigma)}}\,. \tag{3.2.24}$$

We conclude that in order both expressions to agree with one another, K_σ must be 1.

Also, the other correlation functions can be calculated in the same manner, and because θ_{c} is the Josephson phase, and ϕ_{c} the phase of the charge density, we conclude that for $K_\sigma > 1$ the superconductivity is dominant. On the other hand, for $K_\sigma < 1$, the CDW or SDW become dominant. Indeed, because following (3.2.15) the attractive interaction $U < 0$ corresponds to $K_\rho > 1$, and the repulsive interaction $U > 0$ to $K_\rho < 1$, this behaviour is consistent with the physical picture. We conclude that in much the same way as for the Heisenberg model described in Sect. 2.1, the scaling follows the diagonal line, finally reaching the origin.

This is consistent with the fact that the effective Hamiltonian of the repulsive Hubbard model for $U \ll t$ is the Heisenberg model. For the half-filled one-dimensional system, no matter how small $U > 0$ is, we saw that a gap in the charge excitation arises, and we expect that the spin system at energies below this energy gap is described by the Heisenberg model. In a single dimension, because spin-charge separation occurs, and also when the electron filling is so much changed from the half-filled state that the system is no longer an insulator, the spin system is still essentially described by the Heisenberg model.

On the other hand, for the case $U < 0$ of attractive interaction, g_1 becomes relevant, the phase ϕ_{s} of the spin is fixed, and a gap in the spin excitation spectrum arises. Physically, this means that owing to the attractive force between the electrons, spin singlet formation emerges and the system becomes magnetically inert. The charge degree of freedom, on the other hand is gapless. This corresponds to the so-called Luther Emery phase.

We discussed the non-Fermi liquid occurring in the one-dimensional electron system. Now, the question is the relationship between such a one-dimensional electron system and higher-dimensional Fermi liquids. This problem can be tackled using bosonization in higher dimensions. The principal idea is to split the Fermi surface into infinitesimal small segments and to apply one-dimensional bosonization to each segment separately. As has been mentioned at the beginning of this section, these bosons just correspond to the quantized displacement δk_{F} of the Fermi surface. Therefore, Bosonization can be described as an effective theory up to second order of the displacement of the Fermi surface in the k space. In this sense, in the framework of Bosonization, the Fermi surface is considered to be a dynamic object.

This picture almost agrees with the way Fermi liquids are described in Landau's theory. The 'quasi-particle density' $\delta n_{\boldsymbol{k}\sigma}$ occurring there can be interpreted as displacement of the Fermi surface at $\boldsymbol{k}$, and the Landau-parameter $f_{\boldsymbol{k}\sigma,\boldsymbol{k}'\sigma'}$ describing the interaction between the quasi-particles can be seen as a force between different segments of the Fermi surface (just think of a drumhead). Owing to this interaction, the whole Fermi-surface can oscillate in a collective way. These modes are called collective excitation modes. Electron correlation affects these collective exitation modes strongly.

On the other hand, there exist modes that are only contributed by a limited area of the Fermi surface, being local in k-space (individual excitation). These excitations build up a continuum spectrum, and are not very much influenced by the effect of the interaction. Putting it the other way round, also without interaction, individual excitations; that is, electron–hole pair creations emerges. In the higher-dimensional case, because both modes exist and the ratio of the individual excitations is larger, the excitation spectrum does not change drastically when no interaction is present. This corresponds to the Fermi liquid. However, in the one-dimensional case, the 'Fermi surface' consists only of the two points k_{F} and $-k_{\mathrm{F}}$, and therefore only the collective excitation modes exist. Therefore, the effect of the correlation is drastically visible. For this reason, the one-dimensional system is a non-Fermi liquid. We conclude that from the point of view where the Fermi surface is considered to be a dynamic variable, the Tomonaga–Luttinger liquid and the Fermi liquid can be described using almost the same physical picture.

However, in higher dimensions it is not easy to treat the Umklapp scattering by the bosonization scheme. In the Mott insulator, Umklapp scattering certainly occurs, and for this reason, in the vicinity of the Mott insulator state, perhaps a non-Fermi liquid arises. This problem is related to the high-temperature superconductors, and is at present being intensively investigated.

3.3 Magnetic Ordering in Strongly Correlated Electronic Systems

In the discussion of the previous section, we assumed that no long-range order is present, and that no symmetry is broken. In the one-dimensional system, this assumption is justified because strong quantum fluctuations lead to the breakdown of long-range order. However, in two or three dimensions, owing to the interaction, various kinds of long-range orders may occur. In this section, the most important among them, magnetic ordering, will be discussed in the framework of mean field theory.

We consider again the Hubbard model, and express the partition function as path integral

$$Z = \int \mathcal{D}C^\dagger \mathcal{D}C \exp\left[-\int_0^\beta \mathrm{d}\tau\, L\right], \tag{3.3.1}$$

where the Lagrangian is given by

$$L = \sum_{i,\sigma} C_{i\sigma}^\dagger(\partial_\tau - \mu_0)C_{i\sigma} - \sum_{l,l',\sigma} t_{l-l'} C_{l\sigma}^\dagger C_{l'\sigma} + U\sum_i n_{l\uparrow}n_{l\downarrow}\,. \tag{3.3.2}$$

Now, we perform the Stratonovich–Hubbard transformation and obtain

$$Z = \int \mathcal{D}\varphi\mathcal{D}C^\dagger \mathcal{D}C \exp\left[-\int L(\varphi, C^\dagger, C)\,\mathrm{d}\tau\right]. \tag{3.3.3}$$

The Lagrangian then becomes

$$\begin{aligned} L(\varphi, C^\dagger, C) = &\sum_{l,\sigma} C_{l\sigma}^\dagger(\partial_\tau - \mu)C_{l\sigma} - \sum_{l,l',\sigma} t_{l-l'} C_{l\sigma}^\dagger C_{l'\sigma} + \frac{U}{4}\sum_l \varphi_l^2 \\ &+\frac{U}{2}\sum_l \varphi_l(n_{l\uparrow} - n_{l\downarrow})\,. \end{aligned} \tag{3.3.4}$$

Here, we used the rightmost version in (3.2.7) and ignored the interaction $(n_{l\uparrow} + n_{l\downarrow})^2$ between the charges. φ_l is the conjugate field to $n_{l\uparrow} - n_{l\downarrow}$, and can be considered as the magnetic field in this electron system.

In (3.3.4), the interaction between the electrons is reduced to a one-particle problem under the presence of the magnetic field φ_l. However, the price that we have to pay is the functional integration with respect to φ_l. Mean field theory corresponds to the saddle-point approximation with respect to the functional integration in φ_l. That is, after having performed the integration in $C^\dagger$ and C in (3.3.3)

$$Z = \int \mathcal{D}\varphi\, \mathrm{e}^{-A_{\mathrm{eff}}}, \tag{3.3.5}$$

$$A_{\mathrm{eff}} = \frac{U}{4}\int_0^\beta \mathrm{d}\tau \sum_l \varphi_l(\tau)^2 - \mathrm{Tr}\ln\left[\left(\partial_\tau - \mu - t_{l-l'} + \sigma\frac{U}{2}\varphi_l\right)\right], \tag{3.3.6}$$

the extremum $\varphi_l^{\text{saddle}}$ is determined by

$$\delta A_{\text{eff}} = 0\,. \tag{3.3.6$'$}$$

In the mean field approximation, the (imaginary) time-independent solution $\varphi_l^{\text{saddle}}(\tau) = \varphi_l$ is used.

Having introduced the mean field in this way, next we deduce an effective potential for the case when φ_l is small. This corresponds to the Ginzburg–Landau expansion. Writing M for the matrix in the Tr ln term of (3.3.6), its components in the k-space read with $\xi_{\boldsymbol{k}} = \varepsilon(\boldsymbol{k}) - \mu$

$$\begin{aligned}(M)_{(\boldsymbol{k},\mathrm{i}\omega_n,\sigma),(\boldsymbol{k}',\mathrm{i}\omega_n,\sigma')} = \delta_{\sigma\sigma'}\Big[&(-\mathrm{i}\omega_n + \xi_{\boldsymbol{k}})\delta_{\omega_n,\omega_m}\delta_{\boldsymbol{k},\boldsymbol{k}'} \\ &+\frac{1}{\sqrt{\beta N_0}}\frac{\sigma U}{2}\varphi(\boldsymbol{k}-\boldsymbol{k}', \mathrm{i}\omega_n - \mathrm{i}\omega_m)\Big]\,. \end{aligned} \tag{3.3.7}$$

The matrix corresponding to the first term on the right-hand side in (3.3.7) is called $-G_0^{-1}$, the second term is called V. Then, we can perform the expansion

$$\begin{aligned}\operatorname{Tr}\ln M &= \operatorname{Tr}\ln[-G_0^{-1}(1 - G_0V)] \\ &= \operatorname{Tr}\ln[-G_0^{-1}] + \operatorname{Tr}\ln[1 - G_0V] \\ &= \operatorname{Tr}\ln[-G_0^{-1}] - \sum_{n=1}^{\infty}\frac{1}{n}\operatorname{Tr}(G_0V)^n\,. \end{aligned} \tag{3.3.8}$$

We now expand up to second order in V (that is, φ). The first-order term is given by

$$\operatorname{Tr}(G_0V) = \frac{1}{\sqrt{\beta N_0}}\sum_{\mathrm{i}\omega_n,\boldsymbol{k},\sigma} G_0(\mathrm{i}\omega_n,\boldsymbol{k})\frac{\sigma U}{2}\varphi(0,0) = 0 \tag{3.3.9}$$

with $G_0(\mathrm{i}\omega_n,\boldsymbol{k}) = (\mathrm{i}\omega_n - \xi_{\boldsymbol{k}})^{-1}$. The second-order term becomes

$$\begin{aligned}\frac{1}{2}\operatorname{Tr} G_0VG_0V &= \frac{1}{2}\sum_{\substack{\mathrm{i}\omega_n\\ \mathrm{i}\omega_m}}\sum_{\boldsymbol{k},\boldsymbol{k}',\sigma} G_0(\mathrm{i}\omega_n,\boldsymbol{k})(V)_{(\boldsymbol{k},\mathrm{i}\omega_n,\sigma)(\boldsymbol{k}',\mathrm{i}\omega_m,\sigma)} \\ &\quad\times G_0(\mathrm{i}\omega_m,\boldsymbol{k}')(V)_{(\boldsymbol{k}',\mathrm{i}\omega_m,\sigma)(\boldsymbol{k},\mathrm{i}\omega_n,\sigma)} \\ &= \frac{U^2}{4\beta N_0}\sum G_0(\mathrm{i}\omega_n,\boldsymbol{k})\varphi(\mathrm{i}\omega_n - \mathrm{i}\omega_m, \boldsymbol{k}-\boldsymbol{k}') \\ &\quad\times G_0(\mathrm{i}\omega_m,\boldsymbol{k}')\varphi(\mathrm{i}\omega_m - \mathrm{i}\omega_n, \boldsymbol{k}'-\boldsymbol{k}) \\ &= \frac{U^2}{4\beta N_0}\sum_{\boldsymbol{q},\mathrm{i}\omega_l}\left(\sum_{\boldsymbol{k},\mathrm{i}\omega_n} G_0(\mathrm{i}\omega_n,\boldsymbol{k})G_0(\mathrm{i}\omega_n + \mathrm{i}\omega_l, \boldsymbol{k}+\boldsymbol{q})\right) \\ &\quad\times \varphi(\mathrm{i}\omega_l,\boldsymbol{q})\varphi(-\mathrm{i}\omega_l,-\boldsymbol{q})\,. \end{aligned} \tag{3.3.10}$$

Defining the generalized susceptibility $\chi_0(\boldsymbol{q},\mathrm{i}\omega_l)$ as

$$\chi_0(\boldsymbol{q}, \mathrm{i}\omega_l) = \frac{-1}{\beta N_0} \sum_{\boldsymbol{k}, \mathrm{i}\omega_n} G_0(\mathrm{i}\omega_n, \boldsymbol{k}) G_0(\mathrm{i}\omega_n + \mathrm{i}\omega_l, \boldsymbol{k} + \boldsymbol{q}), \tag{3.3.11}$$

the effective action S_{eff} up to second order in φ becomes

$$A_{\text{eff}}^{(2)} = \sum_{\boldsymbol{q}, \mathrm{i}\omega_l} \frac{U}{4} (1 - U\chi_0(\boldsymbol{q}, \mathrm{i}\omega_l)) \varphi(\boldsymbol{q}, \mathrm{i}\omega_l) \varphi(-\boldsymbol{q}, -\mathrm{i}\omega_l). \tag{3.3.12}$$

When performing the sum in $\mathrm{i}\omega_n$ using contour integration, $\chi_0(\boldsymbol{q}, \mathrm{i}\omega_l)$ becomes

$$\chi_0(\boldsymbol{q}, \mathrm{i}\omega_l) = \frac{1}{N_0} \sum_{\boldsymbol{k}} \frac{f(\xi_{\boldsymbol{k}+\boldsymbol{q}}) - f(\xi_{\boldsymbol{k}})}{\mathrm{i}\omega_l + \xi_{\boldsymbol{k}} - \xi_{\boldsymbol{k}+\boldsymbol{q}}}. \tag{3.3.13}$$

Using $\xi_{\boldsymbol{k}} = \xi_{-\boldsymbol{k}}$, after some steps, we obtain

$$\begin{aligned} \chi_0(\boldsymbol{q}, \mathrm{i}\omega_l) &= \frac{1}{N_0} \sum_{\boldsymbol{k}} \frac{f(\xi_{\boldsymbol{k}}) - f(\xi_{\boldsymbol{k}-\boldsymbol{q}})}{\mathrm{i}\omega_l + \xi_{\boldsymbol{k}-\boldsymbol{q}} - \xi_{\boldsymbol{k}}} = \frac{1}{N_0} \sum_{\boldsymbol{k}} \frac{f(\xi_{-\boldsymbol{k}}) - f(\xi_{-\boldsymbol{k}-\boldsymbol{q}})}{\mathrm{i}\omega_l + \xi_{-\boldsymbol{k}-\boldsymbol{q}} - \xi_{-\boldsymbol{k}}} \\ &= \frac{1}{N_0} \sum_{\boldsymbol{k}} \frac{f(\xi_{\boldsymbol{k}}) - f(\xi_{\boldsymbol{k}+\boldsymbol{q}})}{\mathrm{i}\omega_l + \xi_{\boldsymbol{k}+\boldsymbol{q}} - \xi_{\boldsymbol{k}}} = \frac{1}{N_0} \sum_{\boldsymbol{k}} \frac{f(\xi_{\boldsymbol{k}+\boldsymbol{q}}) - f(\xi_{\boldsymbol{k}})}{-\mathrm{i}\omega_l + \xi_{\boldsymbol{k}} - \xi_{\boldsymbol{k}+\boldsymbol{q}}} \\ &= \chi_0(\boldsymbol{q}, -\mathrm{i}\omega_l). \end{aligned} \tag{3.3.14}$$

From the calculation (3.2.14), we also obtain

$$\chi_0(-\boldsymbol{q}, -\mathrm{i}\omega_l) = \chi_0(\boldsymbol{q}, \mathrm{i}\omega_l). \tag{3.3.15}$$

Combining both results, we conclude that $\chi_0(\boldsymbol{q}, \mathrm{i}\omega_l)$ is an even function both in $\boldsymbol{q}$ and in ω_l. We also conclude that $\chi_0(\boldsymbol{q}, \mathrm{i}\omega_l)$ is real, and from

$$\chi_0(\boldsymbol{q}, \mathrm{i}\omega_l) = \frac{1}{N_0} \sum_{\boldsymbol{k}} \frac{(\xi_{\boldsymbol{k}} - \xi_{\boldsymbol{k}+\boldsymbol{q}})(f(\xi_{\boldsymbol{k}+\boldsymbol{q}}) - f(\xi_{\boldsymbol{k}}))}{\omega_l^2 + (\xi_{\boldsymbol{k}} - \xi_{\boldsymbol{k}+\boldsymbol{q}})^2} \tag{3.3.16}$$

we see that $\chi_0(\boldsymbol{q}, \mathrm{i}\omega_l)$ reaches its maximum at $\omega_l = 0$. Therefore, the instability owing to the negative coefficient of the $\varphi\varphi$ term in (3.3.12) is dominated by the $\omega_l = 0$ component. Therefore, in what follows we will only consider this component. Then, the wave number $\boldsymbol{q}$ where $\chi_0(\boldsymbol{q}) = \chi_0(\boldsymbol{q}, \mathrm{i}\omega_l = 0)$ is maximal is determined, and the temperature T_c where this value reaches $1/U$ is the phase transition point with respect to magnetic ordering.

Of course, $\chi_0(\boldsymbol{q})$ depends on the band dispersion $\chi_{\boldsymbol{k}}$, and for $\boldsymbol{q} \to 0$ and $T \to 0$, it can be expressed only in terms of the density of states $D(E_F)$ at the Fermi energy

$$\begin{aligned} \lim_{q \to 0} \chi_0(\boldsymbol{q}) &= \lim_{q \to 0} \frac{1}{N_0} \sum_{\boldsymbol{k}} \frac{f(\xi_{\boldsymbol{k}+\boldsymbol{q}}) - f(\xi_{\boldsymbol{k}})}{\xi_{\boldsymbol{k}} - \xi_{\boldsymbol{k}+\boldsymbol{q}}} = \frac{1}{N_0} \sum_{\boldsymbol{k}} \left[-\frac{\partial f(\xi_{\boldsymbol{k}})}{\partial \xi_{\boldsymbol{k}}} \right] \\ &\xrightarrow[T \to 0]{} \sum_{\boldsymbol{k}} \delta(\xi_{\boldsymbol{k}}) = D(E_F), \end{aligned} \tag{3.3.17}$$

We conclude that for $T = 0$, the condition that ferromagnetism occurs is in this approximation given by

$$D(E_F)U \geqq 1. \tag{3.3.18}$$

This is the so-called Stoner condition.

However, in a realistic material, for the emergence of ferromagnetism, this condition is insufficient. Recalling that $D(E_F) \propto 1/E_F \propto 1/B$ (B: band width), the condition (3.3.18) just becomes the strong-correlation condition $U \geq E_F \sim B$, and it is known that in this case the mean field approximation becomes unreliable. When U becomes large, the electrons try to escape, which leads an effective interaction U_{eff} to be smaller than U. U in (3.3.18) should be replaced by this U_{eff}.

Kanamori determined U_{eff} using the t-matrix approximation. The result is

$$U_{\text{eff}} \sim \frac{U}{1 + U/B} \underset{U \gg B}{\sim} B. \tag{3.3.19}$$

Inserting this into (3.3.18), we obtain

$$D(E_F)B \gtrsim 1. \tag{3.3.20}$$

Because, usually, $D(E_F) \sim B^{-1}$, it is in general difficult to satisfy this condition. Kanamori concluded that ferromagnetism arises when $D(\varepsilon)$ has a broad bandwidth, but at the same time a high peak at the Fermi energy, as is the case, for example, for Ni. However, the conditions necessary for ferromagnetism are deeply related to orbital degeneracy, etc., being a very difficult question, not finally solved at present.

Next, we discuss the problem whether, beside the case $\boldsymbol{q} = \boldsymbol{0}$, magnetic order can occur. In order for $\chi_0(\boldsymbol{q})$ to become large for $\boldsymbol{q} \neq 0$, which conditions might be necessary? From the expression of $\chi_0(\boldsymbol{q})$, we see that it is sufficient that $|f(\chi_{\boldsymbol{k}+\boldsymbol{q}}) - f(\chi_{\boldsymbol{k}})|$ is large and $|\chi_{\boldsymbol{k}} - \chi_{\boldsymbol{k}+\boldsymbol{q}}|$ is small. That is, $\chi_0(\boldsymbol{q})$ is large when electron-hole pairs are created with wave number $\boldsymbol{q}$ having a small excitation energy. In a geometrical sense, this condition means that when the Fermi surface is shifted parallel to the wave vector $\boldsymbol{q}$, it has a large overlap with the original Fermi surface. This is the so-called Nesting condition. When $\boldsymbol{q}$ satisfies this condition, it is called the Nesting vector.

For example, we consider the tight-binding model on a two-dimensional square lattice, and consider the transfer integral only between next-neighbouring states. Then, we have

$$\xi_{\boldsymbol{k}} = -2t(\cos k_x + \cos k_y) - \mu \tag{3.3.21}$$

and $\mu = 0$ corresponds to the half-filled case $\nu = 1$. In this case, as shown in Fig. 3.4, the Fermi surface is a 45-degree declined square. For parallel transport with the vector

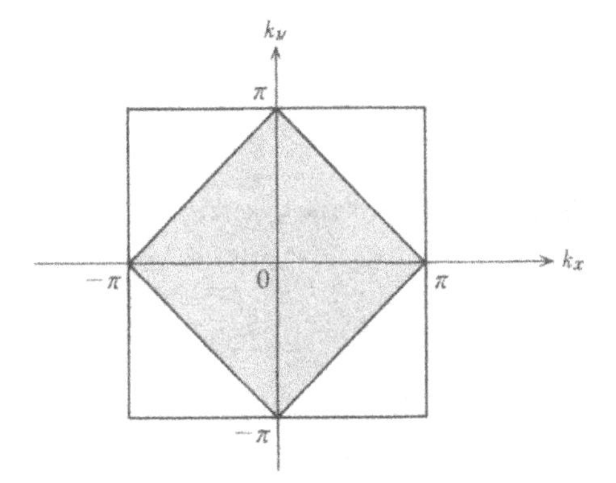

Fig. 3.4. The Fermi surface for the half-filled tight binding model on a two-dimensional rectangular lattice

$$\boldsymbol{Q} = (\pi, \pi)\,, \tag{3.3.22}$$

or the vector related to the reciprocal lattice vector, the Fermi surface is perfectly overlapped. Therefore, magnetic ordering with wave vector $\boldsymbol{Q}$; that is, anti-ferromagnetic ordering, is expected to arise.

Calculating $\chi_0(\boldsymbol{Q})$ explicitly, using the relation $\xi_{\boldsymbol{k}+\boldsymbol{Q}} = -\xi_{\boldsymbol{k}}$, we obtain

$$\begin{aligned}\chi_0(\boldsymbol{Q}) &= \frac{1}{N_0}\sum_{\boldsymbol{k}} \frac{f(-\xi_{\boldsymbol{k}}) - f(\xi_{\boldsymbol{k}})}{2\xi_{\boldsymbol{k}}} = \frac{1}{N_0}\sum_{\boldsymbol{k}} \frac{1}{2\xi_{\boldsymbol{k}}} \tanh\frac{\beta\xi_{\boldsymbol{k}}}{2} \\ &= \int \mathrm{d}\xi \frac{D(\xi)}{2\xi} \tanh\frac{\beta\xi}{2}\,.\end{aligned} \tag{3.3.23}$$

Because in the vicinity of $\xi = 0$, $D(\xi)$ is logarithmically divergent, $D(\xi) \propto \ln t/|\xi|$, we conclude that in the limit $T \to 0$, $\chi_0(\boldsymbol{Q})$ diverges like

$$\chi_0(\boldsymbol{Q}) \sim \left(\ln\frac{t}{T}\right)^2\,. \tag{3.3.24}$$

Therefore, no matter how small U is, there is an instability related to anti-ferromagnetic order.

We now investigate the ordered state using mean field approximation. We insert $\varphi_l = \varphi_0\,\mathrm{e}^{\mathrm{i}\boldsymbol{Q}\boldsymbol{R}_l}$ in (3.3.6), determine $A_{\mathrm{eff}}(\varphi_0)$, and calculate the minimum.

$$\begin{aligned}A_{\mathrm{eff}}(\varphi_0) &= \frac{U}{4}\beta N_0\varphi_0^2 - \sum_{\sigma}\sum_{\omega_n,\boldsymbol{k}:\mathrm{half}} \mathrm{tr}\ln\begin{bmatrix} -\mathrm{i}\omega_n + \xi_{\boldsymbol{k}} & \dfrac{\sigma U\varphi_0}{2} \\ \dfrac{\sigma U\varphi_0}{2} & -\mathrm{i}\omega_n + \xi_{\boldsymbol{k}+\boldsymbol{Q}} \end{bmatrix} \\ &= \frac{U}{4}\beta N_0\varphi_0^2 - 2\sum_{\omega_n,\boldsymbol{k}:\mathrm{half}} \ln\left[-\omega_n^2 - \xi_{\boldsymbol{k}}^2 - \frac{U^2}{4}\varphi_0^2\right]\,.\end{aligned} \tag{3.3.25}$$

Differentiating with respect to φ_0, we obtain

$$\frac{\partial A_{\rm eff}(\varphi_0)}{\partial \varphi_0} = \beta N_0 \left[\frac{1}{2} U \varphi_0 - \frac{1}{\beta N_0} \sum_{\substack{\omega_n \\ k:\text{half}}} \frac{U^2 \varphi_0 / 2}{\omega_n^2 + \xi_k^2 + \dfrac{U^2 \varphi_0^2}{4}} \right] = 0 \,. \quad (3.3.26)$$

This is nothing more than the mean field equation of the gap $\Delta = \frac{1}{2} U \varphi_0$

$$1 = U \frac{1}{N_0} \sum_{k:\text{half}} \frac{1}{E_k} \tanh \frac{\beta E_k}{2} \,. \quad (3.3.27)$$

Here, $E_k = \sqrt{\xi_k^2 + \Delta^2}$ is the band dispersion in the ordered phase.

We conclude that the states in the vicinity of the Fermi surface disappear, and the system becomes an insulator. Owing to the long-range order with the wave number Q, the first Brioullin zone is folded into half, which gives rise to the change from $\nu = 1$ to $\nu = 2$. That is, in this picture anti-ferromagnetic ordering and period doubling are essential for the emerging insulator phase. This point stands in clear contrast to the basic principle of the Mott insulator as discussed in Sect. 3.1. The Mott insulator is even without the magnetic ordering an insulator, and anti-ferromagnetism arises owing to the exchange interaction J. At first sight, both descriptions seems to be very different, however, concerning the anti-ferromagnetic ordered phase, the result is not very different.

To see this, we consider (3.3.27) in the limit $U \gg t$. Because then, the approximation $E_k \simeq \Delta$ ($\varphi_0 \simeq 1/2$) holds, the solution of (3.3.27) is given by

$$\Delta \cong \frac{U}{2} \,. \quad (3.3.28)$$

That is, the energy gap $\sim U$ for the charge transfer is given by the electron-hole pair energy gap 2Δ.

Next, we consider the spin fluctuations. Because φ_l has been fixed for one special z-direction of the spin, it is necessary to find a more general expression. That is, we write the path integral as

$$Z = \int \mathcal{D}\varphi \mathcal{D}\boldsymbol{n} \mathcal{D}C^\dagger \mathcal{D}C \exp\left[- \int L(\varphi, \boldsymbol{n}, C^\dagger, C) \, \mathrm{d}\tau \right] \quad (3.3.29)$$

$$L(\varphi, \boldsymbol{n}, C^\dagger, C) = \frac{U}{4} \sum_l \varphi_l^2 + \sum_{l,\sigma} C_{l\sigma}^\dagger (\partial_\tau - \mu) C_{l\sigma} - \sum_{l,l',\sigma} t_{l-l'} C_{l\sigma}^\dagger C_{l'\sigma} + \frac{U}{2} \sum_{\substack{l \\ \alpha,\beta}} \varphi_l \boldsymbol{n}_l \cdot C_{l\sigma}^\dagger \boldsymbol{\sigma}_{\alpha\beta} C_{l\beta} \,. \quad (3.3.30)$$

Here, we replaced the z-direction in the spin space at every site by some individually different direction $\boldsymbol{n}_l$ ($|\boldsymbol{n}_l|^2 = 1$), and perform the functional integration with respect to $\boldsymbol{n}_l$. Because $\boldsymbol{n}_l$ assigns the direction of the spin, it represents the degree of freedom of the spin wave. Then, φ_l represents in some sense the magnitude of the spin moment and, for $U \gg t$, it will be saturated and become one. In this case, we can replace the φ-integration just by setting $\varphi_l = 1$. (Here, we attach the factor $e^{i\boldsymbol{Q}\boldsymbol{R}_l}$ to $\boldsymbol{n}_l$.)

$$Z \cong \int \mathcal{D}\boldsymbol{n}\mathcal{D}C^\dagger\mathcal{D}C \exp\left[-\int L(\boldsymbol{n}, C^\dagger, C)\,\mathrm{d}\tau\right] \tag{3.3.31}$$

$$\begin{aligned} L(\boldsymbol{n}, C^\dagger, C) &= \sum_l C_l^\dagger(\partial_\tau - \mu)C_l - \sum_{l,l'} t_{l-l'}C_l^\dagger C_{l'} \\ &+ \frac{U}{2}\sum_l C_l^\dagger(\boldsymbol{n}_l \cdot \boldsymbol{\sigma})C_l \,. \end{aligned} \tag{3.3.32}$$

Here, we defined $C_l = [C_{l\uparrow}, C_{l\downarrow}]^t$ and $C_l^\dagger = [C_{l\uparrow}^\dagger, C_{l\downarrow}^\dagger]$.

In mean field theory, $\boldsymbol{n}_l$ is set to be $\boldsymbol{n}_l = \boldsymbol{e}_z\, e^{i\boldsymbol{q}\cdot\boldsymbol{R}_l}$, and the RPA theory determines the influence of small fluctuations around this value. In what follows, a slightly more general discussion will be performed. We represent $\boldsymbol{n}_l$ using $z_{l\sigma}$ ($|z_{l\uparrow}|^2 = |z_{l\downarrow}|^2 = 1$).

$$\boldsymbol{n}_l = \sum_{\alpha,\beta=\uparrow,\downarrow} z_{l\alpha}^* \boldsymbol{\sigma}_{\alpha\beta} z_{l\beta} \,. \tag{3.3.33}$$

We define the corresponding 2×2 matrix as

$$U_l = \begin{bmatrix} z_{l\uparrow} & -z_{l\downarrow}^* \\ z_{l\downarrow} & z_{l\uparrow}^* \end{bmatrix} . \tag{3.3.34}$$

The following identity can easily be proved

$$U_l \sigma^z U_l^\dagger = \boldsymbol{n}_l \cdot \boldsymbol{\sigma} \,. \tag{3.3.35}$$

Defining the new fermions $\tilde{C}_l$ as

$$\tilde{C}_l = U_l^\dagger C_l \,, \qquad \tilde{C}_l^\dagger = C_l^\dagger U_l \tag{3.3.36}$$

we can write (3.3.32) as

$$\begin{aligned} L(U, \tilde{C}^\dagger, \tilde{C}) &= \sum_l \left[\tilde{C}_l^\dagger(\partial_\tau - \mu)\tilde{C}_l + \tilde{C}_l^\dagger U_l^\dagger \partial_\tau U_l \tilde{C}_l\right] \\ &- \sum_{l,l'} t_{l-l'}\tilde{C}_l^\dagger U_l^\dagger U_{l'}\tilde{C}_{l'} + \frac{U}{2}\sum_l \tilde{C}_l^\dagger \sigma^z \tilde{C}_l \,. \end{aligned} \tag{3.3.37}$$

Here, the fermions $\tilde{C}_l^\dagger$ and $\tilde{C}_l$ naturally feel the 'magnetic field' $U/2$ in $-z$ direction, leading to an energetically preferred spin $\downarrow$ direction and an energy cost for the spin $\uparrow$ direction. The density of states corresponding to each state has a bandwidth caused by $t_{l-l'}$. However, in the limit $U \gg t_{l-l'}$, an energy gap of order U arises between both. This gap is called the Hubbard gap, and the high- and log-energy bands are called upper and lower Hubbard bands, respectively. However, we did not assume the ordering of $\boldsymbol{n}_l$, and in this sense this is not the same as the normal band gap.

From the above considerations we conclude that for $U \gg t_{l-l'}$, the state where all spins are assembled in the $\downarrow$ spin band and no spin is in the $\uparrow$ spin state, is a reasonable starting point. That is, writing L as $L_0 + L_{\text{Berry}} + L_t$

$$
\begin{aligned}
L_0 &= \sum_l \tilde{C}_l^\dagger(\partial_\tau - \mu)\tilde{C}_l + \frac{U}{2}\sum_l \tilde{C}_l^\dagger \sigma^z \tilde{C}_l\,, \\
L_{\text{Berry}} &= \sum_l \tilde{C}_l^\dagger U_l \partial_\tau U_l^\dagger \tilde{C}_l\,, \\
L_t &= -\sum_{l,l'} t_{l-l'} \tilde{C}_l^\dagger U_l U_l^\dagger \tilde{C}_{l'}\,,
\end{aligned}
\tag{3.3.38}
$$

we can deduce an effective action in U_l or $\boldsymbol{n}_l$ by performing a perturbative expansion with respect to L_{Berry} and L_t

$$
A_{\text{eff}} = -\operatorname{Tr}\ln\left[\partial_\tau - \mu + \frac{U}{2}\sigma + U_l^\dagger \partial_\tau U_l - t_{l-l'} U_l^\dagger U_{l'}\right]. \tag{3.3.39}
$$

Setting $G_0 = -(\partial_\tau - \mu + U/2\sigma)^{-1}$ and expanding up to first order in ∂_τ and second order in $t_{l-l'}$, we obtain

$$
\begin{aligned}
A_{\text{eff}} &= +\sum_l \operatorname{Tr}\left[G_0(U_l^\dagger \partial_\tau U_l)\right] \\
&\quad + \frac{1}{2}\sum_l \operatorname{Tr}_{l'}\left[G_0 t_{l-l'} U_l^\dagger U_{l'} G_0 t_{l'-l} U_{l'}^\dagger U_l\right].
\end{aligned}
\tag{3.3.40}
$$

The first term on the right-hand side becomes

$$
+\int \mathrm{d}\tau \sum_l \left(U_l^\dagger \partial_\tau U_l\right)_{\downarrow\downarrow} = -\int \mathrm{d}\tau \sum_l \left(z_{l\downarrow}\partial_\tau z_{l\downarrow}^* + z_{l\uparrow}\partial_\tau z_{l\uparrow}^*\right) \tag{3.3.41}
$$

and using polar coordinates φ_l and θ_l for $\boldsymbol{n}_l$,

$$
\begin{aligned}
z_{l\uparrow} &= \cos\frac{\theta_l}{2}\,, \\
z_{l\downarrow} &= \mathrm{e}^{-\mathrm{i}\varphi_l}\sin\frac{\theta_l}{2}\,,
\end{aligned}
\tag{3.3.42}
$$

we obtain

$$
\begin{aligned}
z_{l\downarrow}\partial_\tau z_{l\downarrow}^* + z_{l\uparrow}\partial_\tau z_{l\uparrow}^* &= \mathrm{i}\partial_\tau\varphi_l \cdot \sin^2\frac{\theta_l}{2} \\
&= \frac{\mathrm{i}}{2}\partial_\tau\varphi_l(1-\theta_l)\,.
\end{aligned}
\tag{3.3.43}
$$

Therefore, the Berry phase for $S = 1/2$ given in (2.3.3) and (2.3.4) becomes

$$
\frac{\mathrm{i}}{2}\omega(\{\boldsymbol{n}_l(\tau)\})\,. \tag{3.3.44}
$$

Next, we consider the second term on the right-hand side of (3.3.40)

$$\frac{1}{2}\sum_{l,l'}\frac{1}{\beta}\sum_{\substack{\mathrm{i}\omega_n,\sigma\\ \mathrm{i}\omega_l,\sigma'}}|t_{l-l'}|^2\frac{1}{\mathrm{i}\omega_n-\dfrac{\sigma U}{2}}\cdot\frac{1}{\mathrm{i}\omega_n+\mathrm{i}\omega_l-\dfrac{\sigma' U}{2}}$$
$$\times\left(U_l^\dagger U_{l'}\right)_{\sigma\sigma'}(\mathrm{i}\omega_l)\left(U_{l'}^\dagger U_l\right)_{\sigma\sigma'}(-\mathrm{i}\omega_l)\,. \tag{3.3.45}$$

The summation can be performed as

$$\frac{1}{\beta}\sum_{\mathrm{i}\omega_n}\frac{1}{\left(\mathrm{i}\omega_n-\dfrac{\sigma U}{2}\right)\left(\mathrm{i}\omega_n+\mathrm{i}\omega_l-\dfrac{\sigma' U}{2}\right)}=\frac{f\left(\dfrac{\sigma U}{2}\right)-f\left(\dfrac{\sigma' U}{2}\right)}{\mathrm{i}\omega_l+\dfrac{\sigma-\sigma'}{2}U}\,. \tag{3.3.46}$$

Because the region $|\omega_l| \ll U$ is of interest, we set $\mathrm{i}\omega_l = 0$ and obtain $-(1/U)\delta_{\sigma,\sigma'}$. Then, (3.3.45) becomes

$$\frac{1}{2}\sum_{l,l'}|t_{l-l'}|^2\frac{-1}{U}\int \mathrm{d}\tau\sum_{\sigma}\left(U_l^\dagger U_{l'}\right)_{\sigma,-\sigma}\left(U_{l'}^\dagger U_l\right)_{-\sigma,\sigma}\,. \tag{3.3.47}$$

In an explicit calculation, we obtain

$$\begin{aligned}
&\sum_{\sigma}\left(U_l^\dagger U_{l'}\right)_{\sigma,-\sigma}\left(U_{l'}^\dagger U_l\right)_{-\sigma,\sigma}\\
&=\left(-z_{l\uparrow}^*z_{l'\downarrow}^*+z_{l\downarrow}^*z_{l'\uparrow}^*\right)\left(-z_{l'\downarrow}z_{l\uparrow}+z_{l'\uparrow}z_{l\downarrow}\right)\\
&\quad+\left(-z_{l'\uparrow}^*z_{l\downarrow}^*+z_{l'\downarrow}^*z_{l\uparrow}^*\right)\left(-z_{l\downarrow}z_{l'\uparrow}+z_{l\uparrow}z_{l'\downarrow}\right)\\
&=2z_{l\uparrow}^*z_{l\uparrow}z_{l'\downarrow}^*z_{l'\downarrow}+2z_{l\downarrow}^*z_{l\downarrow}z_{l'\uparrow}^*z_{l'\uparrow}\\
&\quad-2z_{l\uparrow}^*z_{l\downarrow}z_{l'\downarrow}^*z_{l'\uparrow}-2z_{l\downarrow}^*z_{l\uparrow}z_{l'\uparrow}^*z_{l'\downarrow}\\
&=2\left(\frac{1}{2}+S_l^z\right)\left(\frac{1}{2}-S_{l'}^z\right)+2\left(\frac{1}{2}-S_l^z\right)\left(\frac{1}{2}+S_{l'}^z\right)\\
&\quad-2S_l^+S_{l'}^--2S_l^-S_{l'}^+\\
&=1-4\boldsymbol{S}_l\cdot\boldsymbol{S}_{l'}\,,
\end{aligned} \tag{3.3.48}$$

and we conclude that (3.3.47) is given by

$$\int \mathrm{d}\tau\sum_{l,l'}\frac{2|t_{l-l'}|^2}{U}\left(\boldsymbol{S}_l\cdot\boldsymbol{S}_{l'}-\frac{1}{4}\right)\,. \tag{3.3.49}$$

The action consisting of the terms (3.3.44) and (3.3.49) is nothing more than the anti-ferromagnetic Heisenberg model derived in Sect. 3.1.

In such a manner, when the absolute value of the spin moment is saturated in the limit $U \gg t$, the effective action derived for the degree of freedom of the direction $\boldsymbol{n}_l$ leads back to the spin model. Therefore, we conclude that the spin fluctuations around $\boldsymbol{n}_l = \boldsymbol{e}_z\,\mathrm{e}^{\mathrm{i}\boldsymbol{Q}\boldsymbol{R}_l}$ are identical to those of the Heisenberg model. The projection operator that appeared in Sect. 3.1 has been replaced by the exclusion of the spin ↑ component in the rotating frame. In such a way, the Mott insulator can be described in terms of the emerging spin moment, and, in this manner, the mean field state of the anti-ferromagnetic ordering continues to a Mott insulator.

3.4 Self-consistent Renormalization Theory and Quantum Critical Phenomena

As was discussed in the previous section, strong electronic correlation is intimately related to magnetism. Here, we might have the point of view that the magnetic ordering parameter is the most important degree of freedom of the system. From this standpoint, it has been claimed that, it is possible to understand many properties of the normal state as well as those in the ordered state by taking into account the thermal and quantum fluctuations. In particular, there are cases where the magnetic ordering disappears even at zero temperature when some parameter (for example, pressure or carrier density) goes across a critical value. Around these critical values, the quantum fluctuations become large and develop a singularity. These are so-called quantum critical phenomena, leading at low temperature to a non-Fermi-liquid-like behaviour. In this section, the theory of these phenomena is developed.

In the theory of phase transitions, the so-called Landau-Wilson expansion of the free energy is the most fundamental issue. It is obtained using path integral methods by further generalizing the considerations of Sect. 3.3. When the expansion of (3.3.8) is performed up to $n = 4$, the result is

$$A = \frac{1}{2}\sum_{\boldsymbol{q},\mathrm{i}\omega_l} \Pi_2(\boldsymbol{q},\mathrm{i}\omega_l)\varphi(\boldsymbol{q},\mathrm{i}\omega_l)\varphi(-\boldsymbol{q},-\mathrm{i}\omega_l) + \frac{1}{4\beta N_0}\sum_{\substack{\sum_{i=1}^4 \boldsymbol{q}_i=0 \\ \sum_{i=1}^4 \omega_i=0}} \Pi_4(\boldsymbol{q}_i,\mathrm{i}\omega_i)\varphi(\boldsymbol{q}_1,\mathrm{i}\omega_1)\varphi(\boldsymbol{q}_2,\mathrm{i}\omega_2)\varphi(\boldsymbol{q}_3,\mathrm{i}\omega_3)\varphi(\boldsymbol{q}_4,\mathrm{i}\omega_4)\,. \tag{3.4.1}$$

$$\begin{aligned}\Pi_2(\boldsymbol{q},\mathrm{i}\omega_l) &= \frac{U}{2}[1-U\chi_0(\boldsymbol{q},\mathrm{i}\omega_l)]\,,\\ \Pi_4(\boldsymbol{q},\mathrm{i}\omega_l) &= \frac{2(U/2)^4}{\beta N_0}\sum_{\mathrm{i}\varepsilon_n}\sum_{\boldsymbol{k}} G_0(\boldsymbol{k},\mathrm{i}\varepsilon_n)G_0(\boldsymbol{k}+\boldsymbol{k}_1,\mathrm{i}\varepsilon_n+\mathrm{i}\omega_1)\\ &\quad\times G_0(\boldsymbol{k}+\boldsymbol{q}_1+\boldsymbol{q}_2,\mathrm{i}\varepsilon_n+\mathrm{i}\omega_1+\mathrm{i}\omega_2)G_0(\boldsymbol{k}-\boldsymbol{q}_4,\mathrm{i}\varepsilon_n-\mathrm{i}\omega_4)\,.\end{aligned} \tag{3.4.2}$$

When ferromagnetic ordering is considered, the components of $\varphi(\boldsymbol{q},\mathrm{i}\omega_l)$ around $\boldsymbol{q}=\boldsymbol{0}$ are important; in the anti-ferromagnetic case, the components around $\boldsymbol{q}=\boldsymbol{Q}=(\pi,\dots,\pi)=\boldsymbol{G}/2$ are important. Now, we will derive an effective action in the continuum limit assuming that the wave numbers measured from $\mathbf{0}$ or $\boldsymbol{Q}$, and the frequencies, are small.

$\xi_0(\boldsymbol{q}, \mathrm{i}\omega_l)$ defined in (3.3.13) has been discussed for $\mathrm{i}\omega_l = 0$. Now, we will proceed by considering finite $\mathrm{i}\omega_l$. This corresponds to discussing the quantum fluctuations of the spin fluctuation field $\varphi(\boldsymbol{q}, \mathrm{i}\omega_l)$. With $\xi_\pm = \xi_{\boldsymbol{k}\pm\boldsymbol{q}/2}$, we obtain

$$\begin{aligned}\chi_0(\boldsymbol{q}, \mathrm{i}\omega_l) &- \chi_0(\boldsymbol{q}, 0) \\ &= \frac{1}{N_0}\sum_{\boldsymbol{k}} \frac{-\mathrm{i}\omega_l(\xi_+ - \xi_-) + \omega_l^2}{(\xi_+ - \xi_-)[\omega_l^2 + (\xi_+ - \xi_-)^2]}[f(\xi_+) - f(\xi_-)] . \quad (3.4.3)\end{aligned}$$

Owing to the $\boldsymbol{k} \leftrightarrow -\boldsymbol{k}$ symmetry in $\xi_+ - \xi_- = \frac{\boldsymbol{kq}}{m}$, we can write

$$\frac{1}{N_0}\sum_{\boldsymbol{k}} \frac{f(\xi_+) - f(\xi_-)}{\omega_l^2 + (\xi_+ - \xi_-)^2} = 0 . \quad (3.4.4)$$

This can be estimated as $|\xi_+ - \xi_-| \simeq v_\mathrm{F} q$. For $|\omega_l| \leq v_\mathrm{F} q$, the change in $\xi_+ - \xi_-$ can be assumed to be large compared with $|\omega_l|$. Then, we obtain

$$\begin{aligned}&\frac{1}{N_0}\sum_{\boldsymbol{k}} \frac{\omega_l^2}{\omega_l^2 + (\xi_+ - \xi_-)^2} \frac{f(\xi_+) - f(\xi_-)}{\xi_+ - \xi_-} \\ &\simeq -\frac{1}{N_0}\sum_{\boldsymbol{k}} \frac{\omega_l^2}{\omega_l^2 + (\xi_+ - \xi_-)^2}\left[-\frac{\partial f(\xi_+)}{\partial \xi_+}\right] \\ &\simeq -N(\varepsilon_\mathrm{F})\left\langle \frac{\omega_l^2}{\omega_l^2 + (v_\mathrm{F} q)^2 \cos^2\theta}\right\rangle_\mathrm{angle} . \quad (3.4.5)\end{aligned}$$

The bracket $\langle\ \rangle_\mathrm{angle}$ signifies averaging with respect to the angle θ between $\boldsymbol{k}$ and $\boldsymbol{q}$. This calculation can be performed in two and three dimensions, respectively, with the result

$$-cN(\varepsilon_\mathrm{F})\frac{|\omega_l|}{v_\mathrm{F} q} \quad (3.4.6)$$

with $c = 1$ in two dimensions, and $c = \pi/2$ in three dimensions.

From the above calculation, we obtain

$$\Pi_2(\boldsymbol{q}, \mathrm{i}\omega_l) \cong \frac{U}{2}\left[1 - U\chi_0(\boldsymbol{q}, 0) + cUN(\varepsilon_\mathrm{F})\frac{|\omega_l|}{v_\mathrm{F} q}\right] . \quad (3.4.7)$$

The expansion in the ferromagnetic case around $\boldsymbol{q} = \mathbf{0}$ becomes

$$\Pi_2(\boldsymbol{q}, \mathrm{i}\omega_l) \cong \frac{U}{2}\left[(1 - U\chi_0(\mathbf{0}, 0)) + aq^2 + cUN(\varepsilon_\mathrm{F})\frac{|\omega_l|}{v_\mathrm{F} q}\right] . \quad (3.4.8)$$

On the other hand, the expansion in the anti-ferromagnetic case around $\boldsymbol{q} = \boldsymbol{Q}$ becomes

$$\Pi_2(\boldsymbol{q}+\boldsymbol{Q},\mathrm{i}\omega_l) \cong \frac{U}{2}\left[(1-U\chi_0(\boldsymbol{Q},0)) + aq^2 + cUN(\varepsilon_\mathrm{F})\frac{|\omega_l|}{v_\mathrm{F}Q}\right]. \quad (3.4.9)$$

Next, we consider the fourth-order term. In the sense of the renormalization group, higher order terms with respect to $q, \mathrm{i}\omega_l$ in the fourth-order expansion are irrelevant for the critical phenomena. Therefore, we use in the ferromagnetic case the approximation

$$\Pi_4(\boldsymbol{q}_i,\mathrm{i}\omega_i) \to \Pi_4(\boldsymbol{0},0) = \frac{2(U/2)^4}{\beta N_0}\sum_{\mathrm{i}\varepsilon_n}\sum_{\boldsymbol{k}}[G_0(\boldsymbol{k},\mathrm{i}\varepsilon_n)]^4 \quad (3.4.10)$$

and in the anti-ferromagnetic case

$$\begin{aligned}\Pi_4(\boldsymbol{q}_i,\mathrm{i}\omega_i) &\to \Pi_4(\boldsymbol{q}_i=\boldsymbol{Q},0)\\ &= \frac{2(U/2)^4}{\beta N_0}\sum_{\mathrm{i}\varepsilon_n}\sum_{\boldsymbol{k}}[G_0(\boldsymbol{k},\mathrm{i}\varepsilon_n)]^2[G_0(\boldsymbol{k}+\boldsymbol{Q},\mathrm{i}\varepsilon_n)]^2. \quad (3.4.11)\end{aligned}$$

These terms are constants.

After appropriate scaling of space, time and φ coordinate, the above considerations lead to the effective action

$$\begin{aligned}A = &\frac{1}{2}\sum_{\boldsymbol{q},\mathrm{i}\omega_l}\left(\delta_0 + q^2 + \frac{|\omega_l|}{\Gamma_q}\right)\varphi(\boldsymbol{q},\mathrm{i}\omega_l)\varphi(-\boldsymbol{q},-\mathrm{i}\omega_l)\\ &+u\int_0^\beta \mathrm{d}\tau\int \mathrm{d}^d\boldsymbol{r}\,[\varphi(\boldsymbol{r},\tau)]^4. \quad (3.4.12)\end{aligned}$$

The coefficient δ_0 measures the distance from the phase transition point, $\delta_0 < 0$ corresponds to the ordered phase, and $\delta_0 > 0$ to the paramagnetic phase (in the framework where fluctuations are not considered). Here, $\Gamma_q = \Gamma q$ (ferromagnetic case) and $\Gamma_q = \Gamma$ (anti-ferromagnetic case). The relationship between $\varphi(\boldsymbol{r},\tau)$ and $\varphi(\boldsymbol{r},\mathrm{i}\omega_l)$ is given by

$$\varphi(\boldsymbol{r},\tau) = \frac{1}{\sqrt{\beta N_0}}\sum_{\boldsymbol{q},\mathrm{i}\omega_l}\varphi(\boldsymbol{q},\mathrm{i}\omega_l)\,\mathrm{e}^{-\mathrm{i}\omega_l\tau+\mathrm{i}\boldsymbol{q}\cdot\boldsymbol{r}}. \quad (3.4.13)$$

Equation (3.4.12) is the so-called Landau–Wilson expansion of the quantum system. In what follows, several remarks will be made. $\omega_l = 2\pi Tl$ (l: integer) are the bosonic Matsubara frequencies taking discrete values owing to periodic boundary conditions in the imaginary time direction in the interval $0 < \tau < \beta$. This corresponds to the fact that for finite temperature, the system is finite in the imaginary time direction. Next, we consider the coefficient of the second-order term of φ in (3.4.12)

$$D_0^{-1}(\boldsymbol{q},\mathrm{i}\omega_l) = \delta_0 + q^2 + |\omega_l|/\Gamma_q. \quad (3.4.14)$$

q and $|\omega_l|$ appear in a non-symmetric way. Therefore, the system is anisotropic in the space and time directions. As a result, it becomes necessary to introduce a new parameter, the dynamic critical exponent z

$$\omega \sim q^z \tag{3.4.15}$$

as scaling. Equation (3.4.14) becomes symmetric for $z = 3$ (ferromagnetic) and $z = 2$ (anti-ferromagnetic).

As warming up, we consider (3.4.12) for the Gaussian case $u = 0$

$$A_{\text{Gauss}} = \frac{1}{2}\sum_{\boldsymbol{q},\mathrm{i}\omega_l} D_0^{-1}(\boldsymbol{q},\mathrm{i}\omega_l)\varphi(\boldsymbol{q},\mathrm{i}\omega_l)\varphi(-\boldsymbol{q},-\mathrm{i}\omega_l)\,. \tag{3.4.16a}$$

We obtain

$$\begin{aligned} Z_{\text{Gauss}} &= \prod_{\substack{\boldsymbol{q},\mathrm{i}\omega_l:\\ \text{half}}} \int \mathrm{d}\,\mathrm{Re}\,\varphi(\boldsymbol{q},\mathrm{i}\omega_t)\mathrm{d}\,\mathrm{Im}\,\varphi(\boldsymbol{q},\mathrm{i}\omega_l)\,\mathrm{e}^{-D_0^{-1}(\boldsymbol{q},\mathrm{i}\omega_l)[(\mathrm{Re}\,\varphi)^2+(\mathrm{Im}\,\varphi)^2]} \\ &= \prod_{\substack{\boldsymbol{q},\mathrm{i}\omega_l:\\ \text{half}}} [\pi D_0(\boldsymbol{q},\mathrm{i}\omega_l)]\,, \end{aligned} \tag{3.4.16b}$$

and the free energy becomes

$$\begin{aligned} F_{\text{Gauss}} &= -\frac{1}{\beta}\ln Z_{\text{Gauss}} \\ &= -\frac{1}{2\pi}\sum_{\boldsymbol{q},\mathrm{i}\omega_l} \ln \pi D_0(\boldsymbol{q},\mathrm{i}\omega_l) \\ &= \frac{1}{2\beta}\sum_{\boldsymbol{q},\mathrm{i}\omega_l} \ln[\delta_0 + q^2 + |\omega_l|/\Gamma_q] + \text{const.}. \end{aligned} \tag{3.4.16c}$$

Now, we consider the summation in ω_l. Because the expression is analytic at $z = \mathrm{i}\omega_l$, it is possible to write

$$\ln[\delta_0 + q^2 + |\omega_l|/\Gamma_q] = \int_{-\infty}^{\infty} \mathrm{d}\varepsilon \frac{A(\boldsymbol{q},\varepsilon)}{\mathrm{i}\omega_l - \varepsilon}\,. \tag{3.4.17}$$

Here,

$$A(\boldsymbol{q},\varepsilon) = -\frac{1}{\pi}\,\mathrm{Im}\ln\left[\delta_0 + q^2 + \frac{|\omega_l|}{\Gamma_q}\right]\Bigg|_{\mathrm{i}\omega_l\to\varepsilon+\mathrm{i}\delta} \tag{3.4.18}$$

is the imaginary part of the retarded function approaching the real axis from the upper half plane. After some calculations, we obtain

$$A(\boldsymbol{q},\varepsilon) = \frac{1}{\pi}\tan^{-1}\frac{\varepsilon/\Gamma_q}{\delta_0 + q^2}\,. \tag{3.4.19}$$

Owing to $A(\boldsymbol{q},\varepsilon) = -A(\boldsymbol{q},-\varepsilon)$, the right-hand side of (3.4.17) can be expressed as

$$\frac{1}{2}\int_{-\infty}^{\infty} \mathrm{d}\varepsilon\, A(\boldsymbol{q},\varepsilon)\left[\frac{1}{\mathrm{i}\omega_l - \varepsilon} - \frac{1}{\mathrm{i}\omega_l + \varepsilon}\right]\,. \tag{3.4.20}$$

The sum in $\mathrm{i}\omega_l$ can be performed easily for the term in the [ ] brackets (the sum is convergent because for $|\omega_l| \to \infty$ it is of order $\mathcal{O}(\omega_l^{-2})$. Using

$$\frac{1}{2\beta}\sum_{\mathrm{i}\omega_l}\left[\frac{1}{\mathrm{i}\omega_l-\varepsilon}-\frac{1}{\mathrm{i}\omega_l+\varepsilon}\right]=\frac{1}{2}\coth\frac{\varepsilon}{2T} \tag{3.4.21}$$

the free energy (3.4.16c) can be determined. The free energy per unit volume $f_{\mathrm{free}} = F_{\mathrm{free}}/V$ is given by

$$f_{\mathrm{Gauss}}=\frac{1}{N_0}\sum_q\int_0^{\Gamma_q}\frac{\mathrm{d}\varepsilon}{2\pi}\coth\frac{\varepsilon}{2T}\cdot\tan^{-1}\left(\frac{\varepsilon/\Gamma_q}{\delta_0+q^2}\right). \tag{3.4.22}$$

Because equation (3.4.14) is defined in the $|\omega_l|$ plane, $|\varepsilon|$ is limited to $|\varepsilon| < \Gamma_q$. Therefore, the ε integration has a cut off at Γ_q.

Next, we consider the non-linear term. We could think of perturbation theory with respect to u. However, this expansion breaks down for $\delta_0 \to 0$, that is, in the vincinity of the critical point. To see this, let us examine the low-order terms in u. Owing to the non-linearity of A_{int} (3.4.12)

$$Z \cong Z_{\mathrm{Gauss}}\left[1-\langle A_{\mathrm{int}}\rangle_{\mathrm{Gauss}}+\frac{1}{2}\langle A_{\mathrm{int}}^2\rangle_{\mathrm{Gauss}}+\dots\right], \tag{3.4.23}$$

we obtain

$$\langle A_{\mathrm{int}}\rangle_{\mathrm{Gauss}}=u\int\mathrm{d}\tau\,\mathrm{d}^d\boldsymbol{r}\,\left\langle\varphi(\boldsymbol{r},\tau)^4\right\rangle_{\mathrm{Gauss}}=3u\int\mathrm{d}\tau\,\mathrm{d}^d\boldsymbol{r}\,\left[\langle\varphi(\boldsymbol{r},\tau)^2\rangle\right]^2, \tag{3.4.24}$$

and by performing the sum

$$\begin{aligned}\left\langle(\varphi(\boldsymbol{r},\tau))^2\right\rangle_{\mathrm{Gauss}}&=\frac{1}{\beta V}\sum_{\boldsymbol{q},\mathrm{i}\omega_l}\langle\varphi(\boldsymbol{q},\mathrm{i}\omega_l)\varphi(-\boldsymbol{q},-\mathrm{i}\omega_l)\rangle_{\mathrm{Gauss}}\\&=\frac{1}{\beta V}\sum_{\boldsymbol{q},\mathrm{i}\omega_l}D_0(\boldsymbol{q},\mathrm{i}\omega_l)\end{aligned} \tag{3.4.25}$$

in the same way as the sum in $i\omega_l$ calculated above, we obtain

$$\left\langle(\varphi(\boldsymbol{r},\tau))^2\right\rangle_{\mathrm{Gauss}}=\int\frac{\mathrm{d}^d\boldsymbol{q}}{(2\pi)^d}\int_0^{\Gamma_q}\frac{\mathrm{d}\varepsilon}{\pi}\coth\frac{\varepsilon}{2T}\cdot\frac{\varepsilon/\Gamma_q}{(\delta_n+q^2)^2+(\varepsilon/\Gamma_q)^2}, \tag{3.4.26}$$

Now, we discuss the contribution from the region where ε and q are small. Writing $\coth(\varepsilon/2T) \simeq 2T/\varepsilon$, the ε-integration can be performed, with the result

$$\sim T\int\frac{\mathrm{d}^d\boldsymbol{q}}{(2\pi)^d}\frac{1}{\delta_0+q^2}. \tag{3.4.27}$$

This corresponds in (3.4.25) to considering only the term $\mathrm{i}\omega_l = 0$; that is, perturbative expansion in a classical φ^4 model. For $d > 2$, (3.4.27) does not diverge in the limit $\delta_0 \to 0$.

Next, we consider the term in u^2

$$\langle A_{\mathrm{int}}^2 \rangle_{\mathrm{Gauss}} = u^2 \int \mathrm{d}\tau_1\, \mathrm{d}^d\boldsymbol{r}_1\, \mathrm{d}\tau_2\, \mathrm{d}^d\boldsymbol{r}_2 \left\langle \varphi(\boldsymbol{r}_1,\tau_1)^4 \varphi(\boldsymbol{r}_2,\tau_2)^4 \right\rangle_{\mathrm{Gauss}} . \tag{3.4.28}$$

Again, by performing the cumulants expansion of the Gaussian integral, many terms appear. One of them is of the form

$$\int \mathrm{d}\tau_1\, \mathrm{d}^d\boldsymbol{r}_1\, \mathrm{d}\tau_2\, \mathrm{d}^d\boldsymbol{r}_2 \left\langle \varphi(\boldsymbol{r}_1,\tau_1)^2 \right\rangle_{\mathrm{Gauss}} \left\langle \varphi(\boldsymbol{r}_2,\tau_2)^2 \right\rangle_{\mathrm{Gauss}} \times \left\langle \varphi(\boldsymbol{r}_1,\tau_1)\varphi(\boldsymbol{r}_2,\tau_2) \right\rangle_{\mathrm{Gauss}}^2 . \tag{3.4.29}$$

This term contains

$$\begin{aligned}
&\int \mathrm{d}\tau_1\, \mathrm{d}^d\boldsymbol{r}_1 \langle \varphi(\boldsymbol{r}_1,\tau_1)\varphi(\boldsymbol{r}_2,\tau_2)\rangle_{\mathrm{Gauss}}^2 \\
&= \frac{1}{(\beta V)^2} \sum_{\mathrm{i}\omega_l,\boldsymbol{q}} \sum_{\mathrm{i}\omega_l',\boldsymbol{q}'} \langle \varphi(\boldsymbol{q},\mathrm{i}\omega_l)\varphi(-\boldsymbol{q},-\mathrm{i}\omega_l)\rangle_{\mathrm{Gauss}} \\
&\quad \times \langle \varphi(\boldsymbol{q}',\mathrm{i}\omega_l')\varphi(-\boldsymbol{q}',-\mathrm{i}\omega_l')\rangle_{\mathrm{Gauss}} \\
&\quad \times \int_0^\beta \mathrm{d}\tau \int \mathrm{d}^d\boldsymbol{r}\, \mathrm{e}^{\mathrm{i}(\boldsymbol{q}+\boldsymbol{q}')\cdot\boldsymbol{r}}\, \mathrm{e}^{-\mathrm{i}(\omega_l+\omega_l')\tau} \\
&= \frac{1}{\beta V} \sum_{\boldsymbol{q},\mathrm{i}\omega_l} D_0^2(\boldsymbol{q},\mathrm{i}\omega_l) .
\end{aligned} \tag{3.4.30}$$

Let us examine whether or not this term diverges. Performing the sum with respect to $\mathrm{i}\omega_l$, (3.4.30) becomes

$$\int \frac{\mathrm{d}^d\boldsymbol{q}}{(2\pi)^d} \int_0^{\Gamma_q} \frac{\mathrm{d}\varepsilon}{\pi} \coth\frac{\varepsilon}{2T} \cdot \frac{2(\varepsilon/\Gamma_q)(\delta_0+q^2)}{\left[(\delta_0+q^2)^2+(\varepsilon/\Gamma_q)^2\right]^2} , \tag{3.4.31}$$

and again by performing the integration in the region $\varepsilon/2T \ll 1$, the $\mathrm{i}\omega_l = 0$ term of (3.4.30)

$$T \int \frac{\mathrm{d}^d\boldsymbol{q}}{(2\pi)^2} \frac{1}{(\delta_0+q^2)^2} \tag{3.4.32}$$

is obtained. In the case when $d \le 4$, this term is divergent for $\delta_0 \to 0$.

In such a way, close to the critical point, naive perturbation theory breaks down, and it becomes necessary to look for another method to handle all the higher order contributions. The following two methods are representative techniques that have been developed. The first one, developed by Moriya, is called self-consistant renormalization (SCR); the other, which has already been mentioned, is the renormalization group. The principal idea of the SCR-theory is to find the best quadratic effective action that takes into acount the effective, renormalized non-linear term. Moriya developed an elaborated theory and also performed experimental analysis of several materials. In what follows, this will be explained in a simplified manner.

Explicitly, the 'best' action can be determined by the variational method. That is, we choose a quadratic trial action A_0 and choose its parameters such that

$$F = F_0 + \frac{1}{\beta}\langle A - A_0\rangle_{A_0} \tag{3.4.33}$$

becomes minimal. Here, F is related to the true free energy F_{true} by $F_{\text{true}} \leq F$. We choose A_0 to be (3.4.14) with δ_0 replaced by the variation parameter δ:

$$A_0 = \frac{1}{2}\sum_{\boldsymbol{q},\mathrm{i}\omega_l}(\delta + q^2 + |\omega_l|/\Gamma_q)\varphi(\boldsymbol{q},\mathrm{i}\omega_l)\varphi(-\boldsymbol{q},-\mathrm{i}\omega_l)\,. \tag{3.4.34}$$

Then, (3.4.33) becomes

$$\begin{aligned} F = {} & \frac{1}{2\beta}\sum_{\boldsymbol{q},\mathrm{i}\omega_l}\ln[\delta + q^2 + |\omega_l|/\Gamma_q] \\ & + \frac{1}{2\beta}\sum_{\boldsymbol{q},\mathrm{i}\omega_l}\frac{\delta_0 - \delta}{\delta + q^2 + |\omega_l|/\Gamma_q} \\ & + 3u\cdot V\left(\frac{1}{\beta V}\sum_{\boldsymbol{q},\mathrm{i}\omega_l}\frac{1}{\delta + q^2 + |\omega_l|/\Gamma_q}\right)^2 . \end{aligned} \tag{3.4.35}$$

From $\partial F/\partial\delta = 0$, an equation determining δ is obtained:

$$\begin{aligned} \delta &= \delta_0 + 12u\left\langle\varphi(r,\tau)^2\right\rangle_{A_0} = \delta_0 + 12u\frac{1}{\beta V}\sum_{\boldsymbol{q},\mathrm{i}\omega_l}\frac{1}{\delta + q^2 + |\omega_l|/\Gamma_q} \\ &= \delta_0 + 12u\frac{1}{V}\int\frac{\mathrm{d}^d\boldsymbol{q}}{(2\pi)^d}\int_0^{\Gamma_q}\coth\frac{\varepsilon}{2T}\frac{\varepsilon/\Gamma_q}{(\delta + q^2)^2 + (\varepsilon/\Gamma_q)^2}\,. \end{aligned} \tag{3.4.36}$$

δ determined in this manner is related to the correlation length $\xi(T,\delta_0)$, where the non-linear term has been renormalized by

$$\delta(T,\delta_0) = \xi^{-2}(T,\delta_0)\,. \tag{3.4.37}$$

Now, we consider the quantum critical point at zero temperature. This is the point where δ vanishes; δ_0 then obeys the equation

$$0 = \delta_0 + 12u\int\frac{\mathrm{d}^d\boldsymbol{q}}{(2\pi)^d}\int_0^{\Gamma_q}\frac{\mathrm{d}\varepsilon}{\pi}\frac{\varepsilon/\Gamma_q}{q^4 + (\varepsilon/\Gamma_q)^2}\,. \tag{3.4.38}$$

The second term on the right-hand side corresponds to the quantum fluctuations $\langle\varphi(r,\tau)^2\rangle_{T=0}$ at zero temperature. What might be the behaviour of $\delta(T,\delta_0)$ at this value of δ_0, but at finite temperature? To see this, we substract from (3.4.38) equation (3.4.36)

$$\begin{aligned}\delta &= 12u\int\frac{\mathrm{d}^d\boldsymbol{q}}{(2\pi)^d}\int_0^{\Gamma_q}\frac{\mathrm{d}\varepsilon}{\pi}\\ &\quad\times\left\{\coth\frac{\varepsilon}{2T}\frac{\varepsilon/\Gamma_q}{(\delta+q^2)^2+(\varepsilon/\Gamma_q)^2}-\frac{\varepsilon/\Gamma_q}{q^4+(\varepsilon/\Gamma_q)^2}\right\}\\ &= 12u\int\frac{\mathrm{d}^d\boldsymbol{q}}{(2\pi)^d}\int_0^{\Gamma_q}\frac{\mathrm{d}\varepsilon}{\pi}\left(\coth\frac{\varepsilon}{2T}-1\right)\frac{\varepsilon/\Gamma_q}{(\delta+q^2)^2+(\varepsilon/\Gamma_q)^2}\\ &\quad+12u\int\frac{\mathrm{d}^2\boldsymbol{q}}{(2\pi)^d}\int_0^{\Gamma_q}\frac{\mathrm{d}\varepsilon}{\pi}\left\{\frac{\varepsilon/\Gamma_q}{(\delta+q^2)^2+(\varepsilon/\Gamma_q)^2}-\frac{\varepsilon/\Gamma_q}{q^4+(\varepsilon/\Gamma_q)^2}\right\}.\end{aligned} \tag{3.4.39}$$

In what follows, we choose, for example, the case of anti-ferromagnetism ($\Gamma_q=\Gamma$) in $d=3$ dimensions for the analysis of (3.4.39).

For $T\to 0$, as will become clear in a moment, $\delta\ll T$ holds. Therefore, the integral of the first term of (3.4.39) becomes

$$\begin{aligned}&\sim 12u\int\frac{\mathrm{d}^3\boldsymbol{q}}{(2\pi)^3}\int_0^{T}\frac{\mathrm{d}\varepsilon}{2\pi}\frac{2T}{\varepsilon}\frac{\varepsilon/\Gamma}{(\delta+q^2)^2+(\varepsilon/\Gamma)^2}\\ &=\frac{24}{\pi}uT\int\frac{\mathrm{d}^3\boldsymbol{q}}{(2\pi)^3}\frac{1}{\delta+q^2}\tan^{-1}\frac{T/\Gamma}{\delta+q^2}\\ &\simeq\frac{24}{\pi}uT\int\frac{\mathrm{d}^3\boldsymbol{q}}{(2\pi)^3}\frac{1}{q^2}\tan^{-1}\left(\frac{T}{\Gamma_{q^2}}\right)\\ &=\frac{24}{\pi}uT^{3/2}\Gamma^{-1/2}\int\frac{\mathrm{d}^3\boldsymbol{x}}{(2\pi)^3}\frac{1}{x^2}\tan^{-1}\frac{1}{x^2}.\end{aligned} \tag{3.4.40}$$

Here, we substituted the upper boundary of the q integration by $x_\mathrm{c}=T^{-1/2}q_\mathrm{c}$. Because for $x\gg 1$ the integrand behaves like $\propto x^{-4}$, in $d=3$ the integration can be performed with $x_\mathrm{c}\to\infty$. Then, with a being some constant, (3.3.40) finally becomes

$$auT^{3/2}.$$

On the other hand, with $\delta+q^2\ll 1$, the second term of (3.4.39) becomes

$$12\Gamma u\int\frac{\mathrm{d}^3\boldsymbol{q}}{(2\pi)^3}\frac{1}{2\pi}\ln\frac{q^2}{\delta+q^2}=\frac{6\Gamma u}{\pi}\delta^{3/2}\int\frac{\mathrm{d}^3\boldsymbol{x}}{(2\pi)^3}\ln\frac{x^2}{1+x^2}. \tag{3.4.41}$$

In this case, $x_\mathrm{c}\propto\delta^{-1/2}$, and the integrand behaves for $x\gg 1$ like $\propto x^{-2}$, and therefore (3.4.41) becomes of order $\propto u\delta^{3/2}x_\mathrm{c}^{3-2}\propto u\delta^{3/2}\delta^{-1/2}=u\delta$. We conclude that at the quantum critical point, at low temperature, $\delta=\xi^{-2}$ has the following non-trivial temperature dependence:

$$\delta(T) = \xi^{-2}(T) \propto T^{3/2}\,. \tag{3.4.42}$$

In what follows, we will see how this causes critical behaviour of physical quantities.

In the SCR framework, scattering of the electrons by the spin fluctuation field φ causes resistivity. Explicitly, the imaginary part of the self-energy of the electron Green function is the inverse of the mean life time. In lowest order, the self-energy $\Sigma(\boldsymbol{k}, \mathrm{i}\varepsilon_n)$ is given by

$$\Sigma(\boldsymbol{k}, \mathrm{i}\varepsilon_n) = \frac{g^2}{\beta V} \sum_{\boldsymbol{q}, \mathrm{i}\omega_l} G(\boldsymbol{k}+\boldsymbol{q}, \mathrm{i}\varepsilon_n + \mathrm{i}\omega_l) D(\boldsymbol{q}, \mathrm{i}\omega_l)\,. \tag{3.4.43}$$

g is the coupling constant between the spin fluctuations and the electrons. Here, the propagator $D(\boldsymbol{q}, \mathrm{i}\omega_l)$ of φ is given by (3.1.14), where δ_0 has been replaced by the self-consistent value δ. ε_n is the Matsubara frequency of the electrons

$$G(\boldsymbol{k}, \mathrm{i}\varepsilon_n) = \frac{1}{\mathrm{i}\varepsilon_n - \xi_{\boldsymbol{k}}}\,. \tag{3.4.44}$$

Performing the sum over ω_l in (3.4.43), we obtain

$$\Sigma(\boldsymbol{k}, \mathrm{i}\varepsilon_n) = \frac{g^2}{V} \sum_{\boldsymbol{q}} \int_{-\Gamma_q}^{\Gamma_q} \frac{\mathrm{d}\varepsilon}{\pi} \frac{n(\varepsilon) + f(\xi_{\boldsymbol{k}+\boldsymbol{q}})}{\mathrm{i}\varepsilon_n + \varepsilon - \xi_{\boldsymbol{k}+\boldsymbol{q}}} \frac{\varepsilon/\Gamma_q}{(\delta + q^2)^2 + (\varepsilon/\Gamma_q)^2}\,. \tag{3.4.45}$$

Here, $n(\varepsilon) = (\mathrm{e}^{\beta\varepsilon} - 1)^{-1}$ is the bose distribution. By analytic continuation $\mathrm{i}\varepsilon_n \to \omega + \mathrm{i}\delta$, we obtain the retarded self-energy $\Sigma^{\mathrm{R}}(\boldsymbol{k}, \omega)$

$$\begin{aligned} \operatorname{Im} \Sigma^{\mathrm{R}}(\boldsymbol{k}, \omega) = \frac{-\pi g^2}{V} \sum_{\boldsymbol{q}} \int_{-\Gamma_q}^{\Gamma_q} \frac{\mathrm{d}\varepsilon}{\pi} \frac{\varepsilon/\Gamma_q}{(\delta + q^2)^2 + (\varepsilon/\Gamma_q)^2} \\ \times [n(\varepsilon) + f(\omega + \varepsilon)] \delta(\omega + \varepsilon - \xi_{\boldsymbol{k}+\boldsymbol{q}})\,. \end{aligned} \tag{3.4.46}$$

Using $n(\varepsilon) + f(\varepsilon) = 1$ at zero temperature, we see that $\operatorname{Im} \Sigma^{\mathrm{R}}(\boldsymbol{k}, \omega = 0)$ vanishes at $T = 0$.

At finite temperature, because $n(\varepsilon) + f(\varepsilon) \simeq T/\varepsilon$ for $|\varepsilon| \ll T$, we obtain

$$\operatorname{Im} \Sigma^{\mathrm{R}}(k_{\mathrm{F}}, 0) \cong -\frac{\pi g^2}{V} \sum_{\boldsymbol{q}} \int_{-\Gamma_q}^{\Gamma_q} \frac{\mathrm{d}\varepsilon}{\pi} \frac{T/\Gamma_q}{(\delta + q^2)^2 + (\varepsilon/\Gamma_q)^2} \delta(\varepsilon - \xi_{\boldsymbol{k}+\boldsymbol{q}})\,. \tag{3.4.47}$$

Considering again the anti-ferromagnetic case in $d = 3$ dimensions, owing to $\Gamma_q = \Gamma \gg \delta$, $\Gamma \gg q^2$, approximatively $\xi_{\boldsymbol{k}+\boldsymbol{q}} \simeq \boldsymbol{v}_{\mathrm{F}} \cdot \boldsymbol{q}$, and (3.4.47) becomes

$$\operatorname{Im} \Sigma^{\mathrm{R}}(k_{\mathrm{F}}, 0) \cong -\frac{g^2}{V} \sum_{\boldsymbol{q}} \frac{T/\Gamma}{(\delta + q^2)^2 + (\xi_{\boldsymbol{k}+\boldsymbol{q}}/\Gamma)^2}\,. \tag{3.4.48}$$

The dominant contribution of the $\boldsymbol{q}$ integral comes from the region

$$\delta + q^2 \sim \frac{\xi_{\boldsymbol{k}+\boldsymbol{q}}}{\Gamma} \sim \frac{v_{\mathrm{F}}}{\Gamma} q ,$$

therefore the approximation $\delta + q^2 \simeq \delta$ holds, leading to

$$-\operatorname{Im} \Sigma^{\mathrm{R}}(k_{\mathrm{F}}, 0) \cong g^2 \int_0^{T/v_{\mathrm{F}}} \frac{\mathrm{d}^d \boldsymbol{q}}{(2\pi)^d} \frac{T/\Gamma}{\delta^2 + (\boldsymbol{v}_{\mathrm{F}} \cdot \boldsymbol{q}/\Gamma)^2} . \tag{3.4.49}$$

In the three-dimensional case, averaging over the angle leads to

$$\begin{aligned} \left\langle \frac{1}{\delta^2 + (v_{\mathrm{F}} q \cos\theta/\Gamma)^2} \right\rangle &= \frac{1}{2} \int_{-1}^{1} \mathrm{d}(\cos\theta) \frac{1}{\delta^2 + ((v_{\mathrm{F}} q/\Gamma)\cos\theta)^2} \\ &\cong \frac{1}{2} \int_{-\infty}^{\infty} \mathrm{d}x \frac{1}{\delta^2 + (v_{\mathrm{F}} q/\Gamma x)^2} = \frac{\pi}{2} \frac{\Gamma}{v_{\mathrm{F}} q \delta} . \end{aligned} \tag{3.4.50}$$

As a result, (3.4.49) becomes

$$\begin{aligned} -\operatorname{Im} \Sigma^{\mathrm{R}}(k_{\mathrm{F}}, 0) &\cong \int_0^{T/v_{\mathrm{F}}} \frac{\mathrm{d}^3 \boldsymbol{q}}{(2\pi)^3} \frac{\pi g^2}{2} \frac{T}{v_{\mathrm{F}} q \cdot \delta} \\ &\sim g^2 \left(\frac{T}{v_{\mathrm{F}}} \right)^3 \frac{1}{\delta} \sim T^{3/2} . \end{aligned} \tag{3.4.51}$$

Compared with normal T^2 dependence, this is enhanced at low temperatures. This can be interpreted as the resistivity of the system becoming larger owing to large spin fluctuations close to the quantum critical point.

In such a way, other physical quantities – specific heat, susceptibility, nuclear spin relaxation rate – can also be determined in arbitrary dimensions. For more details, the reader is referred to the publications of Moriya [8]. One characteristic of the SCR theory is that the temperature dependence of the magnetic susceptibility is induced by the temperature dependence of the correlation length ξ and the magnitude of the spin moments. For example, in the case of weak ferromagnetism, in a large temperature interval the Curie law $1/T$ is obtained, but the physical picture is different from the case of localized spins, where the spin fluctuation is totally uncorrelated in coordinate space and the spin moment is fixed at s.

The SCR theory presented above could be called the quantum version of the theory of mode coupling describing a phase transition. Correspondingly, there should also exist a quantum version of the renormalization group. This will be pointed out in what follows. We go back again to the Gaussian theory (3.4.14). Its free energy is given in (3.4.22) in terms of an integral over the wave number $\boldsymbol{q}$ and the energy ε. Now, we will change the integration region little by little and discuss the modifications.

First, we substitute the cut-off Λ of the $\boldsymbol{q}$ integral by λ/b ($b > 1$, where $b - 1$ is assumed to be infinitesimally small).

$$
\begin{aligned}
f_{\text{Gauss}} &= \left(\int_0^{\Lambda/b} + \int_{\Lambda/b}^{\Lambda} \right) \frac{\mathrm{d}^d \boldsymbol{q}}{(2\pi)^d} \int_0^{\Gamma_q} \frac{\mathrm{d}\varepsilon}{2\pi} \coth \frac{\varepsilon}{2T} \tan^{-1} \frac{\varepsilon/\Gamma_q}{\delta_0 + q^2} \\
&= \int_0^{\Lambda/b} (\cdots) + \Lambda^d K_d \ln b \int_0^{\Gamma_\Lambda} \frac{\mathrm{d}\varepsilon}{2\pi} \coth \frac{\varepsilon}{2T} \tan^{-1} \frac{\varepsilon/\Gamma_\Lambda}{\delta_0 + \Lambda^2} \\
&\equiv f'_{\text{Gauss}} + f_\Lambda \ln b \,.
\end{aligned}
\tag{3.4.52}
$$

Here, $K_d = \int \mathrm{d}^d \boldsymbol{k}/(2\pi)^d \delta(k-1)$ is $1/(2\pi)^d$ times the surface of the d-dimensional unit ball. The first term on the right-hand side is transformed into a form similar to the original one under the scale transformation

$$
\begin{aligned}
\boldsymbol{q} &= \boldsymbol{q}'/b \qquad (\boldsymbol{r} = \boldsymbol{r}'b) \,, \\
\delta_0 &= \delta'_0/b^2 \,, \\
\varepsilon &= \varepsilon'/b^z \qquad (\tau = \tau' b^z) \,, \\
T &= T'/b^z \,,
\end{aligned}
\tag{3.4.53}
$$

that is,

$$
\begin{aligned}
f'_{\text{Gauss}} = b^{-(d+z)} &\int_0^{\Lambda} \frac{\mathrm{d}^d \boldsymbol{q}'}{(2\pi)^d} \int_0^{\Gamma_{(q'b^{-1})(n)} b^z} \frac{\mathrm{d}\varepsilon'}{2\pi} \\
&\times \coth \left(\frac{\varepsilon'}{2T'} \right) \tan^{-1} \left[\frac{\varepsilon'/\Gamma_{(q'b^{-1})}}{\delta'_0 + q'^2} \cdot b^{2-z} \right] ,
\end{aligned}
\tag{3.4.54}
$$

where z is defined by

$$
\Gamma_{(q'b^{-1})} b^{z-2} = \Gamma_{q'} \,. \tag{3.4.55}
$$

Then, for ferromagnetism ($\Gamma_q = \Gamma q$), z is given by $z = 3$, and for anti-ferromagnetism, ($\Gamma_q = \Gamma$), z equals $z = 2$. Equation (3.4.54) becomes

$$
f'_{\text{Gauss}} = b^{-(d+z)} \int_0^{\Gamma} \frac{\mathrm{d}^d \boldsymbol{q}'}{(2\pi)^d} \int_0^{\Gamma_{q'} b^2} \frac{\mathrm{d}\varepsilon'}{2\pi} \coth \left(\frac{\varepsilon'}{2T'} \right) \tan^{-1} \left[\frac{\varepsilon'/\Gamma_{q'}}{\delta'_0 + q'^2} \right]
\tag{3.4.56}
$$

and we obtain for f'_{Gauss}

$$
\begin{aligned}
f'_{\text{Gauss}} &= b^{-(d+z)} \int_0^{\Lambda} \frac{\mathrm{d}^d \boldsymbol{q}'}{(2\pi)^d} \int_0^{\Gamma_{q'}} \frac{\mathrm{d}\varepsilon'}{2\pi} (\cdots) \\
&\quad + b^{-(d+z)} \int_0^{\Lambda} \frac{\mathrm{d}^d \boldsymbol{q}'}{(2\pi)^d} \coth \left(\frac{\Gamma_{q'}}{2T} \right) \tan^{-1} \left[\frac{1}{\delta'_0 + q'^2} \right] \cdot 2 \frac{\Gamma_{q'}}{2\pi} \ln b \\
&\equiv f''_{\text{Gauss}} + f_\Gamma \ln b \,.
\end{aligned}
\tag{3.4.57}
$$

Up to this point, we have split the integration region in the $q-\varepsilon$ plane into the two regions indicated in Fig. 3.5, where the contribution of the oblique region is given by $(f_\Lambda + f_\Gamma) \ln b$. To determine how the parameters contained in the theory change, when the cut off is made smaller and smaller, and

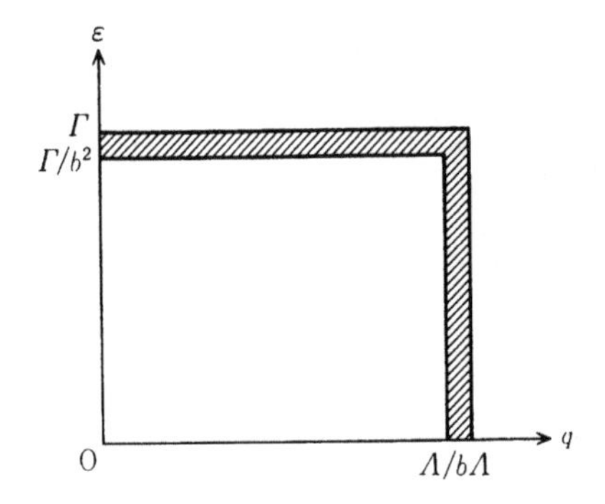

Fig. 3.5. The renormalization process in the $q\varepsilon$ plane for the anti-ferromagnetic case. The oblique region is integrated out

the pursuit of this flow, is called the method of the renormalization group (Sect. 2.1). That is, the requirement that f''_{Gauss} has the same form after the scaling (3.4.53) leads to the scaling equations

$$\begin{aligned}\frac{\mathrm{d}\delta_0(b)}{d\ln b} &= 2\delta_0(b)\,,\\ \frac{\mathrm{d}T(b)}{d\ln b} &= zT(b)\,.\end{aligned}\tag{3.4.58}$$

In the framework of path integrals, the renormalization group method proceeds within the following three steps.

(1) The component of φ corresponding to the oblique region in Fig. 3.5 (notated in the following $\partial\Lambda$) called φ_1, and the component φ_0 of the remaining inner region are decomposed as $\varphi = \varphi_0 + \varphi_1$. An effective action for φ_0 is determined by integrating out the φ_1-component.
(2) By rescaling q and ε, the cut off Λ is set to its original value.
(3) φ_0 is rescaled so that the coefficient of q^2 again becomes $1/2$ (wave function renormalization).

The first step is the most difficult. We will perform the integration by using perturbation theory with respect to u. The zeroth order in u is the Gaussian case that has already be discussed

$$\begin{aligned}\mathrm{e}^{-A_{\mathrm{eff}}(\varphi_0)} &= \int \mathcal{D}\varphi_1\,\mathrm{e}^{-A(\varphi_0+\varphi_1)}\\ &= \int \mathcal{D}\varphi_1\,\mathrm{e}^{-A_{\mathrm{Gauss}}(\varphi_0)-A_{\mathrm{Gauss}}(\varphi_1)-A_{\mathrm{int}}(\varphi+\varphi_1)}\\ &= \mathrm{e}^{-A_{\mathrm{Gauss}}(\varphi_0)}Z_1\left\langle \mathrm{e}^{-A_{\mathrm{int}}(\varphi_0+\varphi_1)}\right\rangle_{\varphi_1,\mathrm{Gauss}}\end{aligned}\tag{3.4.59}$$

where Z_1 is given by

$$\begin{aligned}Z_1 &= \int \mathcal{D}\varphi_1\,\mathrm{e}^{-A_{\mathrm{Gauss}}(\varphi_1)}\\ \langle\mathcal{O}\rangle_{\varphi_1,\mathrm{Gauss}} &= \frac{1}{Z_1}\int \mathcal{D}\varphi_1\mathcal{O}\,\mathrm{e}^{-A_{\mathrm{Gauss}}(\varphi_1)}\,.\end{aligned}\tag{3.4.60}$$

Expanding (3.4.59) with respect to A_{int}, we obtain (in what follows, $\langle\ \rangle_{\varphi_1,\text{Gauss}}$ will be notated as $\langle\ \rangle_{\varphi_1}$)

$$\begin{aligned}\left\langle \mathrm{e}^{-A_{\text{int}}(\varphi_0+\varphi_1)}\right\rangle_{\varphi_1} &= 1 - \langle A_{\text{int}}\rangle_{\varphi_1} + \tfrac{1}{2}\langle A_{\text{int}}^2\rangle_{\varphi_1} - \dots \\ &\approx \exp\left[-\langle A_{\text{int}}\rangle_{\varphi_1} + \tfrac{1}{2}\left\{\langle A_{\text{int}}^2\rangle_{\varphi_1} - \langle A_{\text{int}}\rangle_{\varphi_1}^2\right\}\right] . \qquad (3.4.61)\end{aligned}$$

This expansion will now be evaluated.

$$\begin{aligned}\langle A_{\text{int}}\rangle_{\varphi_1} &= u\int \mathrm{d}^d\boldsymbol{r}\,\mathrm{d}\tau\,\left\langle(\varphi_0+\varphi_1)^4\right\rangle_{\varphi_1} \\ &= u\int \mathrm{d}^d\boldsymbol{r}\,\mathrm{d}\tau\,\left[\varphi_0^4 + 6\varphi_0^2\langle\varphi_1^2\rangle_{\varphi_1} + \langle\varphi_1^4\rangle_{\varphi_1}\right] . \qquad (3.4.62)\end{aligned}$$

Here, $\langle\varphi_1^4\rangle_{\varphi_1}$ is a constant not containing φ_0. The term $\langle\varphi_1^2\rangle_{\varphi_1}$ that is proportional to φ_0^2 is given by

$$\begin{aligned}\langle\varphi_1^2\rangle_{\varphi_1} &= \int_{\partial\Lambda}\frac{\mathrm{d}^d\boldsymbol{q}}{(2\pi)^d}\frac{\mathrm{d}\varepsilon}{\pi}\coth\frac{\varepsilon}{2T}\frac{\varepsilon/\Gamma_q}{(\delta_0+q^2)^2+(\varepsilon/\Gamma_q)^2} \\ &= \Lambda^d K_d \ln b\int_0^{\Gamma_\Lambda}\frac{\mathrm{d}\varepsilon}{\pi}\coth\frac{\varepsilon}{2T}\cdot\frac{\varepsilon/\Gamma_\Lambda}{(\delta_0+\Lambda)^2+(\varepsilon/\Gamma_\Lambda)^2} \\ &\quad + \frac{2\ln b}{\pi}\int_0^{\Lambda}\frac{\mathrm{d}^d\boldsymbol{q}}{(2\pi)^d}\coth\frac{\Gamma_q}{2T}\cdot\frac{1}{(\delta_0+q^2)^2+1} . \qquad (3.4.63)\end{aligned}$$

Setting $\Lambda = \Gamma_\Lambda = 1$ in what follows, we obtain

$$\langle\varphi_1^2\rangle_{\varphi_1} = f^{(2)}(T,\delta_0)\ln b \qquad (3.4.64)$$

$$\begin{aligned}f^{(2)}(T,\delta_0) &= \frac{2}{\pi}\int_0^1\frac{\mathrm{d}^d\boldsymbol{q}}{(2\pi)^d}\coth\frac{q^{z-2}}{2T}\cdot\frac{1}{(\delta_0+q^2)^2+1} \\ &\quad + K_d\int_0^1\frac{\mathrm{d}\varepsilon}{\pi}\coth\frac{\varepsilon}{2T}\cdot\frac{\varepsilon}{(\delta_0+1)^2+\varepsilon^2} . \qquad (3.4.65)\end{aligned}$$

Next, the A_{int}^2 term is given by (using the short-cut notation $\mathrm{d}1 \equiv \mathrm{d}r_1\,\mathrm{d}\tau_1$)

$$\begin{aligned}&\frac{1}{2}u^2\int \mathrm{d}1,\mathrm{d}2\left[\left\langle(\varphi_0(1)+\varphi_1(1))^4(\varphi_0(2)+\varphi_1(2))^4\right\rangle_{\varphi_1}\right. \\ &\quad \left. -\left\langle(\varphi_0(1)+\varphi_1(1))^4\right\rangle_{\varphi_1}\left\langle(\varphi_0(2)+\varphi_1(2))^4\right\rangle_{\varphi_1}\right] . \qquad (3.4.66)\end{aligned}$$

Here, the φ_0^2 term is already a first-order correction in u, which will be ignored. We focus on the correction of the φ_0^4 term, given by

$$\begin{aligned}&\frac{1}{2}u^2\int \mathrm{d}1\,\mathrm{d}2\cdot 36\varphi_0^2(1)\varphi_0^2(2)\left[\langle\varphi_1^2(1)\varphi_1^2(2)\rangle_{\varphi_1} - \langle\varphi_1^2(1)\rangle_{\varphi_1}\langle\varphi_1^2(2)\rangle_{\varphi_1}\right] \\ &\equiv u^2\int \mathrm{d}1\,\mathrm{d}2\,G_{\varphi_1^2}(1-2)\varphi_0^2(1)\varphi_0^2(2) . \qquad (3.4.67)\end{aligned}$$

Here, because φ_0^2 varies slowly compared with the fast oscillating $G_{\varphi_1^2}(1-2)$ component, (3.4.67) becomes approximately

$$(3.4.67) \cong 18 \int \mathrm{d}(1-2) G_{\varphi_1^2}(1-2) \int \mathrm{d}1\, \varphi_0^4(1)\,. \tag{3.4.68}$$

The integration gives

$$\begin{aligned}
&\int \mathrm{d}(1-2) G_{\varphi_1^2}(1-2) \\
&\quad = 2 \int \mathrm{d}(1-2) \langle \varphi_1(1)\varphi_1(2)\rangle_{\varphi_1^2} \\
&\quad = 2 \int \mathrm{d}(1-2) [D_0(1-2)]^2 \\
&\quad = 2 \int \mathrm{d}(1-2) \frac{1}{\beta V} \sum_{\boldsymbol{q}, \mathrm{i}\omega_l \in \partial\Lambda} \frac{1}{\beta V} \sum_{\boldsymbol{q}', \mathrm{i}\omega_m \in \partial\Lambda} \mathrm{e}^{\mathrm{i}(\boldsymbol{q}+\boldsymbol{q}')\cdot(\boldsymbol{r}_1-\boldsymbol{r}_2)} \\
&\qquad \times \mathrm{e}^{-\mathrm{i}(\omega_l+\omega_m)(\tau_1-\tau_2)} D_0(\boldsymbol{q}, \mathrm{i}\omega_l) D_0(\boldsymbol{q}', \mathrm{i}\omega_m) \\
&\quad = 2 \cdot \frac{1}{\beta V} \sum_{\boldsymbol{q}, \mathrm{i}\omega_l \in \partial\Lambda} D_0(\boldsymbol{q}, \mathrm{i}\omega_l) D_0(-\boldsymbol{q}, -\mathrm{i}\omega_l) \\
&\quad = 2 \cdot \frac{1}{\beta V} \sum_{\boldsymbol{q}, \mathrm{i}\omega_l \in \partial\Lambda} \frac{1}{(\delta_0 + q^2 + |\omega_l|/\Gamma_q)^2} \\
&\quad = 2 \int_{\partial\Lambda} \frac{\mathrm{d}^d \boldsymbol{q}}{(2\pi)^d} \int \frac{\mathrm{d}\varepsilon}{2\pi} \coth\frac{\varepsilon}{2T} \cdot \frac{(\varepsilon/\Gamma_q)(\delta_0+q^2)}{[(\delta_0+q^2)^2 + (\varepsilon/\Gamma_q)^2]^2} \equiv 2 f^{(4)} \ln b\,.
\end{aligned} \tag{3.4.69}$$

Here, setting again $\Lambda = \Gamma_\Lambda = 1$, $f^{(4)}$ is given by

$$\begin{aligned}
f^{(4)} = &\frac{2}{\pi} \int_0^1 \frac{\mathrm{d}^d \boldsymbol{q}}{(2\pi)^d} \coth\frac{q^{z-2}}{2T} \cdot \frac{q^{z-2}(\delta_0+q^2)}{[(\delta_0+q^2)^2+1]^2} \\
&+ K_d \int_0^1 \frac{\mathrm{d}\varepsilon}{\pi} \coth\frac{\varepsilon}{2T} \cdot \frac{\varepsilon(\delta_0+1)}{[(\delta_0+1)^2+\varepsilon^2]^2}\,.
\end{aligned} \tag{3.4.70}$$

Putting the pieces together, the effective action of φ_0 is given by

$$\begin{aligned}
A_{\text{eff}}(\varphi_0) = &\frac{1}{2} \sum_{\substack{\boldsymbol{q}, \mathrm{i}\omega_l \\ \in \Lambda - \partial\Lambda}} \left(\delta_0 + 12u f^{(2)} \ln b + q^2 + \frac{|\omega_l|}{\Gamma_q}\right) \varphi_0(\boldsymbol{q}, \mathrm{i}\omega_l) \varphi_0(-\boldsymbol{q}, -\mathrm{i}\omega_l) \\
&+ [u - 18u^2 f^{(4)} \ln b] \int \mathrm{d}^d \boldsymbol{r}\, \mathrm{d}\tau\, \varphi_0^4(\boldsymbol{r}, \tau)\,.
\end{aligned} \tag{3.4.71}$$

Next, we perform the wave function renormalization, in order that the coefficient of $q^2\varphi^2$ is invariant, we impose for φ_0 the scaling transformation

$$\varphi_0(\boldsymbol{q}, \mathrm{i}\omega_l) = b^{(d+z+2)/2}\varphi'(\boldsymbol{q}', \mathrm{i}\omega_l')$$
$$\varphi_0(\boldsymbol{r}, \tau) = b^{(-d-z+2)/2}\varphi'(\boldsymbol{r}', \tau'). \tag{3.4.72}$$

Doing so, (3.4.71) becomes

$$A'_{\text{eff}}(\varphi') = \frac{1}{2}\int \frac{\mathrm{d}^d\boldsymbol{q}'}{(2\pi)^d}\sum_{\mathrm{i}\omega_l}\left(\delta_0[1+2\ln b] + 12uf^{(2)}\ln b + q'^2 + \frac{|\omega_l'|}{\Gamma_{q'}}\right)$$
$$\times\,\varphi'(\boldsymbol{q}', \mathrm{i}\omega_l')\varphi'(-\boldsymbol{q}', -\mathrm{i}\omega_l') \tag{3.4.73}$$
$$+(u(1+(4-d-z)\ln b) - 18u^2 f^{(4)}\ln b]\int \mathrm{d}r'\,\mathrm{d}\tau'\,\varphi^4(\boldsymbol{r}', \tau').$$

From this equation, we can read off the transformation of δ_0 and u. Because the scaling law (3.4.58) of T remains unchanged, the renormalization group equations are given by

$$\frac{\mathrm{d}T(b)}{\mathrm{d}\ln b} = zT(b), \tag{3.4.74a}$$

$$\frac{\mathrm{d}\delta_0(b)}{\mathrm{d}\ln b} = 2\delta_0(b) + 12u(b)f^{(2)}, \tag{3.4.74b}$$

$$\frac{\mathrm{d}u(b)}{\mathrm{d}\ln b} = [4-(d+z)]u(b) - 18u(b)^2 f^{(4)}. \tag{3.4.74c}$$

Also, without determining the explicit solution of these equations, many conclusions can be drawn. First, we discuss the relationship with the classical limit. We notice that the quantum system can be considered as a $d+1$-dimensional classical system, with finite size β in the original imaginary time direction. Owing to the anisotropy in the space and time directions, the dynamical critical exponent z has been introduced in (3.4.15) by $\omega \sim T \sim q^z \sim \xi^{-z}$. This z appears in (3.4.74a). That is, from (3.4.74a), we obtain

$$T(b) = Tb^z. \tag{3.4.75}$$

When large scales (large b) are considered, at the point where $T(b)$ becomes $T(b) = 1$, the system starts to feel the finite size in the imaginary time dimension. We notate b_0 for this b and obtain $b_0 \simeq T^{-1/z}$.

When b becomes even larger ($b > b_0$), the properties of the system will become similar to a d-dimensional classical system. This dimensional cross-over corresponds to a quantum/classical cross-over. When ξ is the correlation length of the system, this cross-over will occur at $b \simeq \xi$. Following (3.4.74), for $T(b) > 1$, $f^{(2)}, f^{(4)}$ behave like $f^{(2)}$, $f^{(4)} \propto T(b)$, and considering $v(b) \equiv u(b)T(b)$ instead of $u(b)$ as the coefficient of φ^4, (3.4.74) becomes

$$\frac{\mathrm{d}\delta_0(b)}{\mathrm{d}\ln b} = 2\delta_0(b) + Cv(b), \tag{3.4.76a}$$

$$\frac{\mathrm{d}v(b)}{\mathrm{d}\ln b} = (4-d)v(b) - Dv(b)^2. \tag{3.4.76b}$$

For $T(b) \gg 1$, $f^{(2)}$ and $f^{(4)}$ have been set $f^{(2)} = CT(b)$ and $f^{(4)} = DT(b)$. These equations are nothing more than the usual renormalization group equations of the φ^4 theory. Comparing (3.4.74) with (3.4.76), we conclude that the condition whether $u(b)$ and $v(b)$ are relevant/irrelevant is only shifted by z. That is, the upper critical dimension d_u for the quantum critical phenomen (in all dimensions larger than d_u, the non-linear interaction becomes irrelevant, and the Gauss approximation is the correct description of the critical phenomen) is given by $d_u + z = 4$. As has been mentioned above, z is given by 2 or 3, therefore, in $d = 3$, d is usually larger than d_u. Then, the critical exponent is given by the theory of Landau. In what follows, this case will be considered further; however, when the temperature is increased, the system changes and is described by (3.4.76), and non-classical critical phenomena described by Wilson's theory and the corresponding critical exponents will appear. This cross-over will now be analysed.

First, when the $u(b)^2$ term in (3.4.74c) is ignored, we obtain

$$u(b) = ub^{4-(d+z)} . \tag{3.4.77}$$

Inserting this into (3.4.74b), we obtain

$$\delta_0(b) = b^2 \left[\delta_0 + 12u \int_0^{\ln b} \mathrm{d}x \, \mathrm{e}^{[2-(d+z)]x} f^{(2)}(T\,\mathrm{e}^{zx}) \right] . \tag{3.4.78}$$

Writing b_1 for the value of b with $\delta(b) \simeq 1$, we obtain $T(b_1) \ll 1$ when $1 \ll b_1 \ll b_0$, and in this case at $T = 0\,\mathrm{K}$ quantum critical phenomena appear. We will explicitly determine this condition. Replacing $f^{(2)}(T\,\mathrm{e}^{zx})$ in (3.4.78) by $f^{(2)}(0)$, we obtain

$$\delta_0(b) = b^2 \left[\delta_0 + \frac{12uf^{(2)}(T=0)}{z+d-2} \right] \equiv b^2 r . \tag{3.4.79}$$

Here, r measures the distance from the critical point for $T = 0\,\mathrm{K}$. Because $b_1 = r^{-1/2}$ is obtained, inserting this in (3.4.75), we obtain

$$T(b_1) = Tr^{-z/2} \ll 1 . \tag{3.4.80}$$

That is, the $T = 0\,\mathrm{K}$ quantum critical phenomena will occur at temperatures as low as $T \ll r^{z/2}$.

On the other hand, on the high-temperature side $T \gg r^{z/2}$, because $b_0 \ll b_1$, the integration region in b must be split into the two parts $1 < b < b_0$ and $b_0 < b < b_1$. Because for $1 < b < b_0$, condition (3.4.77) can be applied, we obtain

$$v(b_0) = T(b_0)u(b_0) = u(b_0) = ub_0^{4-(d+z)} = uT^{(d+z-4)/z} . \tag{3.4.81}$$

Furthermore, by splitting the integration of $f(T\,\mathrm{e}^{zx})$ in (3.4.78) in the zero temperature part $f(T = 0)$ and $f(T\,\mathrm{e}^{zx}) - f(T = 0)$, we obtain instead of (3.4.79) the finite temperature corrected equation

$$\delta_0(b_0) = T^{-2/z}\left[r + BuT^{(d+z-2)/z}\right]. \tag{3.4.82}$$

The term in [ ] brackets equals ξ^{-2}, where ξ is the correlation length of the Gaussian theory. Notice that in the anti-ferromagnetic case ($z = 2$) in $d = 3$, the result becomes $\xi^{-2} \propto r + \text{const} \times \mathrm{T}^{3/2}$ and coincides with the result of the SCR theory described above. Here, B is a constant given by

$$B = 12\frac{1}{z}\int_0^1 \mathrm{d}T \cdot T^{[2-(d+2z)]/z}[f^{(2)}(T) - f^{(2)}(0)]. \tag{3.4.83}$$

Using $v(b_0)$ and $\delta_0(b_0)$ as initial values, we perform the $b_0 < b < b_1$ integration in (3.4.76). If $v(b_1) \ll 1$, then it is justified to consider the Gaussian theory. This condition is called the Ginzburg criterion. This will now be analysed further. We set in what follows $d = 3$. Then we obtain from (3.4.76b)

$$v(b) = v(b_0)\,\mathrm{e}^{\ln(b/b_0)}. \tag{3.4.84}$$

Here, for $b < b_1$, we get by assumption $v(b) \ll 1$, and therefore the $v(b)^2$ term has been ignored. Inserting this solution into (3.4.76a), the result is

$$\delta_0(b) = (\delta_0(b_0) + Cv(b_0))\,\mathrm{e}^{2\ln(b/b_0)} - Cv(b_0)\,\mathrm{e}^{\ln(b/b_0)}. \tag{3.4.85}$$

With this equation, $\delta(b_1) = 0$ can be solved, with the result

$$\frac{b_1}{b_0} \cong [\delta_0(b_0) + Cv(b_0)]^{-1/2} = T^{1/z}\left[r + (B + C)uT^{1+1/z}\right]^{-1/2}, \tag{3.4.86}$$

where

$$v(b_1) = uT\left[r + (B + c)uT^{1+1/z}\right] \ll 1 \tag{3.4.87}$$

is the Ginzberg criterion. On the other hand, for $v(b_1) \gg 1$, the system is in the Wilson region. The derivation of the non-classical critical exponents in this critical region is given in many textbooks, to which the interested reader is referred. The above considerations can be summarized as shown in Fig. 3.6.

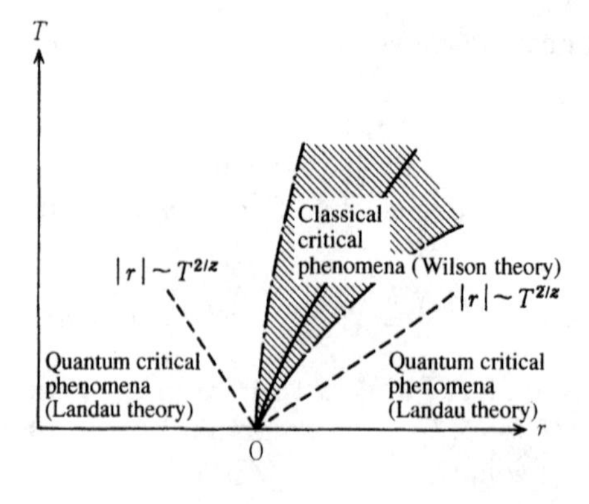

Fig. 3.6. Schematic outline of the regions where quantum critical phenomena occur [J.A. Hertz: Phys. Rev. B14 (1976) 1165]

In such a way, in quantum critical phenomena, where quantum fluctuations play a crucial role for mediating the phase transition, many new features appear that do not arise in classical systems. Quantum critical phenomena have been observed in anti-ferromagnetic ordered heavy electron systems and in dielectric substances showing quantum paradielectricity, and are the subject of present research. However, in cases where it is difficult to define an order parameter, as is the case for the Mott insulator, the understanding of quantum critical phenomena is still poor. This problem is still unsolved. Furthermore, we should not forget about the fact that in realistic systems, scattering at impurities cannot be ignored, and that there is the possibility that owing to the long-range character of the Coulomb force, the phase transition becomes first order.

4. Local Electron Correlation

In this chapter, local electron correlation is discussed. Roughly speaking, this problem has been well understood in one-dimensional systems. It is closely related and gives many implications to the other strongly correlated electronic systems. First, the problem of magnetically active impurities in metals – the Kondo problem – will be examined. Next, the dynamical mean field theory in the limit of large space dimension d will be tackled as a further development of the Kondo problem.

4.1 The Kondo Effect

In Sect. 3.1, many different models of electronic correlation were discussed. Here, we will focus on the problem of magnetically active impurities in the metal. We start by analysing qualitatively the physics contained in (3.1.22). Using the Stratonovich-Hubbard transformation as was done in Sect. 3.3, the Lagrangian becomes

$$\begin{aligned} L = &\sum_{\boldsymbol{k},\sigma} C^\dagger_{\boldsymbol{k}\sigma}(\partial_\tau - \mu + \varepsilon_{\boldsymbol{k}})C_{\boldsymbol{k},\sigma} + \sum_\sigma f^\dagger_\sigma(\partial_\tau + E_\mathrm{f})f_\sigma \\ &+ \frac{U}{4}\varphi(\tau)^2 + \frac{U}{2}\varphi(\tau)(f^\dagger_\uparrow f_\uparrow - f^\dagger_\downarrow f_\downarrow) \\ &- \frac{1}{\sqrt{N_0}}\sum_{\boldsymbol{k},\sigma}\left(V_{\boldsymbol{k}}C^\dagger_{\boldsymbol{k},\sigma}f_\sigma + V^*_{\boldsymbol{k}}f^\dagger_\sigma C_{\boldsymbol{k}\sigma}\right). \end{aligned} \tag{4.1.1}$$

where φ is the spin fluctuation field. First, we integrate out $C^\dagger$ and C and determine the effective theory of φ, $f^\dagger$ and f:

$$\begin{aligned} A_\mathrm{eff}(\{\varphi\}, f^\dagger, f) = &\sum_{\mathrm{i}\omega_n,\sigma}(-\mathrm{i}\omega_n + E_\mathrm{f} + \Sigma_\mathrm{f}(\mathrm{i}\omega_n))f^\dagger_\sigma(\mathrm{i}\omega_n)f_\sigma(\mathrm{i}\omega_n) \\ &+ \frac{U}{2}\int \mathrm{d}\tau \sum_\sigma \sigma\varphi(\tau)f^\dagger_\sigma(\tau)f_\sigma(\tau) + \frac{U}{4}\int \mathrm{d}\tau\, \varphi(\tau)^2. \end{aligned} \tag{4.1.2}$$

The free energy of the f electron is given by

$$\Sigma_{\mathrm{f}}(\mathrm{i}\omega_n) = \frac{1}{N_0}\sum_{\boldsymbol{k}} \frac{|V_{\boldsymbol{k}}|^2}{\mathrm{i}\omega_n - \xi_{\boldsymbol{k}}} . \tag{4.1.3}$$

Splitting this expression into the real and imaginary parts, under the assumption that $|\omega_n|$ is smaller than the band width by replacing $|V_{\boldsymbol{k}}|^2$ with $\langle|V_{\boldsymbol{k}}|^2\rangle$ we obtain

$$\begin{aligned}\Sigma_{\mathrm{f}}(\mathrm{i}\omega_n) &= \left\langle|V_{\boldsymbol{k}}|^2\right\rangle_{E_{\mathrm{F}}} \int \mathrm{d}\xi\, D(\xi) \frac{-\mathrm{i}\omega_n - \xi_{\boldsymbol{k}}}{\omega_n^2 + \xi_{\boldsymbol{k}}^2} \\ &\cong -\mathrm{i}\pi D(E_{\mathrm{F}}) \left\langle|V_{\boldsymbol{k}}|^2\right\rangle_{E_{\mathrm{F}}} \operatorname{sgn}\omega_n + \operatorname{Re}\Sigma_{\mathrm{f}}(\mathrm{i}\omega_n) \\ &\equiv -\mathrm{i}\Delta_0 \operatorname{sgn}\omega_n + \operatorname{Re}\Sigma_{\mathrm{f}}(\mathrm{i}\omega_n) .\end{aligned} \tag{4.1.4}$$

Redefining E_{f} by adsorbing $\operatorname{Re}\Sigma_{\mathrm{f}}(\mathrm{i}\omega \simeq 0)$, (4.1.2) becomes

$$\begin{aligned}A_{\mathrm{eff}}\left(\{\varphi\}, f^{\dagger}, f\right) &= \sum_{\mathrm{i}\omega_n,\sigma} (-\mathrm{i}\omega_n + E_{\mathrm{f}} - \mathrm{i}\Delta_0 \operatorname{sgn}\omega_0) f_{\sigma}^{\dagger}(\mathrm{i}\omega_n) f_{\sigma}(\mathrm{i}\omega_n) \\ &\quad + \frac{U}{2}\int \mathrm{d}\tau \sum_{\sigma} \sigma\varphi(\tau) f_{\sigma}^{\dagger}(\tau) f_{\sigma}(\tau) \\ &\quad + \frac{U}{4}\int \mathrm{d}\tau\, \varphi(\tau)^2 .\end{aligned} \tag{4.1.5}$$

We now determine the action of the static component of φ. This corresponds to the mean field theory as first introduced by Anderson:

$$A(\varphi) = -\sum_{\mathrm{i}\omega_n,\sigma} \ln\left(-\mathrm{i}\omega_n + E_{\mathrm{f}} - \mathrm{i}\Delta_0 \operatorname{sgn}\omega_n + \frac{U}{2}\sigma\varphi\right) + \frac{U}{4}\beta\varphi^2 . \tag{4.1.6}$$

Differentiating with respect to φ, we obtain

$$\begin{aligned}&\beta^{-1}\frac{\mathrm{d}A(\varphi)}{\mathrm{d}\varphi} \\ &= -\frac{1}{\beta}\sum_{\mathrm{i}\omega_n,\sigma}\left[\frac{U\sigma/2}{-\mathrm{i}\omega_n + E_{\mathrm{f}} - \mathrm{i}\Delta_0 \operatorname{sgn}\omega_n + U\varphi\sigma/2} + U\varphi/2\right] \\ &= \frac{1}{\beta}\sum_{\mathrm{i}\omega_n}\left[\frac{U/2}{\mathrm{i}\omega_n + \mathrm{i}\Delta_0 \operatorname{sgn}\omega_n - E_{\mathrm{f}} - U\varphi/2}\right. \\ &\quad \left. - \frac{U/2}{\mathrm{i}\omega_n + \mathrm{i}\Delta_0 \operatorname{sgn}\omega_n - E_{\mathrm{f}} + U\varphi/2}\right] + \frac{U}{2}\varphi \\ &= \frac{U}{2}\left[\frac{1}{\beta}\sum_{\mathrm{i}\omega_n} \frac{U\varphi}{(\mathrm{i}\omega_n + \mathrm{i}\Delta_0 \operatorname{sgn}\omega_n - E_{\mathrm{f}})^2 - (U\varphi/2)^2} + \varphi\right] .\end{aligned} \tag{4.1.7}$$

Performing the limit $T \to 0$, we obtain

$$\begin{aligned}\beta^{-1}\frac{\mathrm{d}A(\varphi)}{\mathrm{d}\varphi} &= -\frac{U^2}{2}\varphi\int_0^\infty \frac{\mathrm{d}\omega}{2\pi}\left\{\frac{1}{(\omega+\Delta_0+\mathrm{i}E_\mathrm{f})^2+(U\varphi/2)^2}\right.\\ &\quad\left.+\frac{1}{(\omega+\Delta_0-\mathrm{i}E_\mathrm{f})^2+(U\varphi/2)^2}\right\}+\frac{U\varphi}{2}\\ &= -\frac{U}{2\pi}\left[\tan^{-1}\frac{\omega+\Delta_0+\mathrm{i}E_\mathrm{f}}{U\varphi/2}+\tan^{-1}\frac{\omega+\Delta_0-\mathrm{i}E_\mathrm{f}}{U\varphi/2}\right]_{\omega=0}^{\omega=\infty}+\frac{U\varphi}{2}\\ &= -\frac{U}{2\pi}\left\{\tan^{-1}\frac{U\varphi/2}{\Delta_0+\mathrm{i}E_\mathrm{f}}+\tan^{-1}\frac{U\varphi/2}{\Delta_0-\mathrm{i}E_\mathrm{f}}\right\}+\frac{U\varphi}{2}\,. \end{aligned} \tag{4.1.8}$$

The integral has been calculated using the formula

$$\int \mathrm{d}x\tan^{-1}\frac{x}{a} = x\tan^{-1}\frac{x}{a}-\frac{a}{2}\ln(a^2+x^2)\,. \tag{4.1.9}$$

In order to determine the qualitative behaviour of $A(\varphi)$, it is sufficient to consider only the small components of φ.

Up to first order in φ, we obtain

$$\frac{\mathrm{d}A(\varphi)}{\mathrm{d}\varphi} = \frac{U}{2}\left[1-\frac{1}{\pi}\frac{\Delta_0 U}{\Delta_0^2+E_\mathrm{f}^2}\right]\varphi\,. \tag{4.1.10}$$

When this coefficient is negative, the action $A(\varphi)$ has a minimum at a finite value of φ; correspondingly, in the mean field limit, a local moment emerges. This is the so-called Anderson criterion.

This can be understood as follows. Δ_0 is the inverse of the life time of the f-electron, due to the hybridization $V_{\boldsymbol{k}}$ with the conduction electrons. On the other hand, the farther away the energy E_f is from the Fermi surface $\xi = 0$ of the conduction band, the more difficult is the single electron occupancy of the f electron, therefore

$$\Delta = \frac{\Delta_0^2+E_\mathrm{f}^2}{\Delta_0} \tag{4.1.11}$$

determines the energy scale that obstructs the occurrence of a localized moment. This energy scale competes with the repulsive Coulomb force U; for $U > \pi\Delta$, a localized moment is supposed to emerge, and for $U < \pi\Delta$, it will not emerge. Especially for the case $U \gg \pi\Delta$, the spin moment of the f-electron is almost saturated to the value 1/2, and a localized spin moment will appear. This limit is the so-called Kondo limit. In this case, following (4.1.2), in the Green function of the f electron peaks at $\omega = E_\mathrm{f} \pm U/2$ with weight 1/2 appear, respectively.

However, the above discussion has been limited to classical φ, that is, the degree of the static mode. So, what do we expect when quantum fluctuations of φ are included in the discussion? This is the heart of the problem of local-electron correlation, i.e., the Kondo problem. In a direct manner, the quantum fluctuations can be understood as tunnelling between the two minima $+\varphi_0$ and $-\varphi_0$ of $A(\varphi)$. At a characteristic frequency ω_0, a cross-over occurs between two physical pictures. For $\omega \gg \omega_0$, or $T \gg \omega_0$, the observation time is shorter than the inverse of the transition frequency between $\pm\varphi_0$, and therefore φ can be assumed to stay fixed at one place $+\varphi_0$ or $-\varphi_0$. Therefore, the localized spin moment should be observable, and for example the susceptibility obeys the Curie law ($S = 1/2$)

$$\chi(T) = \frac{S(S+1)\mu_{\mathrm{B}}^2}{3T} . \tag{4.1.12}$$

On the other hand, for $\omega \ll \omega_0$, or $T \ll \omega_0$, the observation time is longer than ω_0^{-1}, the spin moment disappears because in this time scale, φ often switches between $+\varphi_0$ and $-\varphi_0$. ω_0 measured in units of temperature is called the Kondo temperature T_{K}, and around T_{K}, the cross-over between these two physical picture happens. For example, the susceptibility (4.1.12) is saturated at $T \simeq T_{\mathrm{K}}$ and reaches a constant value

$$\chi(T \to 0) \sim \frac{S(S+1)\mu_{\mathrm{B}}^2}{T_{\mathrm{K}}} \tag{4.1.13}$$

at lower temperatures.

The discussion of this characteristic temperature T_K is the next problem, and we will tackle it using some kind of mean field theory. In order to do so, we generalize (3.1.22) slightly and write

$$\begin{aligned} H = &\sum_{m=1}^{N}\sum_{k}(\varepsilon_k - \mu)C_{km}^{\dagger}C_{km} + E_{\mathrm{f}}\sum_{m=1}^{N} f_m^{\dagger}f_m + U\left(\sum_{m=1}^{N} f_m^{\dagger}f_m\right)^2 \\ &-\sqrt{\frac{2}{N_0 N}}\sum_{m=1,k}^{N} V_k\left(C_{km}^{\dagger}f_m + f_m^{\dagger}C_{km}\right) . \end{aligned} \tag{4.1.14}$$

The spin label σ in (4.1.1) has been replaced by m, running from $1, \ldots, N$. For example, when M orbits are present, we set $N = 2M$. In what follows, we analyse the case $N \gg 1$. This corresponds to the so-called $1/N$ expansion.

We now introduce the slave-boson method. This is a technique to exclude the double occupation of f electrons in the limit $U \to \infty$. We allow for the f-electron only the $N + 1$ states

$$|\mathrm{vac}\rangle, |m\rangle = f_m^{\dagger}|\mathrm{vac}\rangle \qquad (m = 1 \sim N) \tag{4.1.15}$$

with $|\mathrm{vac}\rangle$ being the state with no f electron. The next goal is to construct a description with different operators leading to equivalent matrix elements in this $(N+1)$-dimensional Hilbert space. Explicitly, we consider a fermionic and a bosonic space, and identify vectors that are elements of its direct product space with (4.1.15) as

$$|\text{vac}\rangle \leftrightarrow b^\dagger|0,0\rangle\,,$$
$$|m\rangle = f_m^\dagger|\text{vac}\rangle \leftrightarrow s_m^\dagger|0,0\rangle\,. \tag{4.1.16}$$

Here, we introduced one boson and N different fermions, and $|0,0\rangle$ is the vacuum state with no boson and no fermion. The boson $b^\dagger, b$ that has been introduced to describe the unoccupied-fermion state is called the slave boson. Under the identification (4.1.16), with the relation

$$f_m \leftrightarrow s_m b^\dagger, \quad f_m^\dagger \leftrightarrow s_m^\dagger b \tag{4.1.17}$$

between the operators, all matrix elements become equivalent. For example,

$$\langle\text{vac}|f_m|m\rangle = \langle\text{vac}|f_m f_m^\dagger|\text{vac}\rangle = 1 \tag{4.1.18}$$

corresponds to

$$\langle 0,0|b s_m b^\dagger s_m^\dagger|0,0\rangle = \langle 0,0|b b^\dagger s_m s_m^\dagger|0,0\rangle = 1\,. \tag{4.1.19}$$

Owing to the introduction of the slave bosons, the constraint condition originally expressed as the inequality

$$\sum_m f_m^\dagger f_m \le 1 \tag{4.1.20}$$

is now given by the equation

$$b^\dagger b + \sum_m s_m^\dagger s_m = 1\,. \tag{4.1.21}$$

This is the main advantage of this method, because (4.1.21) can easily be included in the path integral using Lagrange multipliers.

After these preliminaries, we can express the Lagrangian (4.1.14) in this new language by

$$\begin{aligned} L = {} & \sum_{m=1}^{N}\sum_{\boldsymbol{k}}(\partial_\tau + \xi_{\boldsymbol{k}})C_{\boldsymbol{k}m}^\dagger C_{\boldsymbol{k}m} + \sum_{m=1}^{N} s_m^\dagger(\partial_\tau + E_{\mathrm{f}})s_m + b^\dagger\partial_\tau b \\ & -\sqrt{\frac{2}{N_0 N}}\sum_{m=1}^{N}\sum_{\boldsymbol{k}} V_{\boldsymbol{k}}\left(C_{\boldsymbol{k}m}^\dagger s_m b^\dagger + s_m^\dagger b C_{\boldsymbol{k}m}\right) \\ & +\lambda\left(\sum_m s_m^\dagger s_m + b^\dagger b - 1\right)\,. \end{aligned} \tag{4.1.22}$$

Here, instead of U we introduced the field λ, being the Lagrange multiplier field to express the constraint.

Looking at the Lagrangian (4.1.22), it becomes clear that it is 'almost' proportional to N. We achieve this goal by using the following technique.

First, we require that when $\sum_{m=1}^{N} s_m^\dagger s_m$ is of order N, then also $b^\dagger b$ is of order N. Writing $b = \sqrt{N} b_0$ and $b^\dagger = \sqrt{N} b_0^\dagger$, (4.1.21) becomes

$$N b_0^\dagger b_0 + \sum_{m=1}^{N} s_m^\dagger s_m = Nq\,. \tag{4.1.23}$$

After the calculation, we set $q = 1/N$. Then, L is of order N, and in the limit $N \to \infty$, the saddle-point method is accurate. Integrating out the fermions $C^\dagger$ and C, we obtain an effective Lagrangian for $b_0^\dagger$, b_0 and λ:

$$\begin{aligned} A_{\mathrm{eff}}\left(b_0^\dagger, b_0, \lambda\right) &= -N \operatorname{Tr} \ln \begin{bmatrix} \partial_\tau + \xi_{\boldsymbol{k}} & -\sqrt{\dfrac{2}{N_0}} V_{\boldsymbol{k}} b_0^\dagger \\ -\sqrt{\dfrac{2}{N_0}} V_{\boldsymbol{k}} b_0 & \partial_\tau + E_{\mathrm{f}} + \lambda \end{bmatrix} \\ &\quad + N \int \mathrm{d}\tau \left(b_0^\dagger (\partial_\tau + \lambda) b_0 - \lambda q \right). \end{aligned} \tag{4.1.24}$$

In the limit $N \to \infty$, the integral can be replaced by the saddle-point solution

$$\delta A_{\mathrm{eff}} = 0\,. \tag{4.1.25}$$

When the saddle-point solution is determined for time-independent $b_0^\dagger$, b_0 and λ, the fermion integral is just given by (4.1.1) for $U = 0$ with the substitutions

$$V_{\boldsymbol{k}} \to b_0^\dagger V_{\boldsymbol{k}}\,, \quad E_{\mathrm{f}} \to E_{\mathrm{f}} + \lambda\,. \tag{4.1.26}$$

Therefore, corresponding to (4.1.6), we obtain

$$\begin{aligned} A_{\mathrm{eff}}\left(b_0^\dagger, b_0, \lambda\right) &= -N \sum_{\mathrm{i}\omega_n} \ln(-\mathrm{i}\omega_n + E_{\mathrm{f}} + \lambda - \mathrm{i}|b_0|^2 \Delta_0 \operatorname{sgn} \omega_n) \\ &\quad + \beta N \lambda (|b_0|^2 - q)\,. \end{aligned} \tag{4.1.27}$$

Here, the constraint (4.1.23) corresponds to

$$\begin{aligned} (N\beta)^{-1} \frac{\partial A_{\mathrm{eff}}}{\partial \lambda} &= -\frac{1}{\beta} \sum_{\mathrm{i}\omega_n} \frac{\mathrm{e}^{\mathrm{i}\omega_n \delta}}{-\mathrm{i}\omega_n + E_{\mathrm{f}} + \lambda - \mathrm{i}|b_0|^2 \Lambda_0 \operatorname{sgn} \omega_n} + |b_0|^2 - q \\ &= 0\,. \end{aligned} \tag{4.1.28}$$

The first term on the right-hand side corresponds to $\langle s_m^\dagger s_m \rangle$, where we added the factor $\mathrm{e}^{\mathrm{i}\omega_n \delta}$ because of $\tau = -\delta$ (δ: infinitesimal) in the Green function.

On the other hand, $b_0^\dagger$ and b_0 appear only in the combination $|b_0|^2$

$$\begin{aligned} (N\beta)^{-1} \frac{\partial A_{\mathrm{eff}}}{\partial (|b_0|^2)} &= +\frac{1}{\beta} \sum_{\mathrm{i}\omega_n, \sigma} \frac{\mathrm{i}\Delta_0 \operatorname{sgn} \omega_n}{-\mathrm{i}\omega_n + E_{\mathrm{f}} + \lambda - \mathrm{i}|b_0|^2 \Delta_0 \operatorname{sgn} \omega_n} + \lambda \\ &= 0\,, \end{aligned} \tag{4.1.29}$$

and this equation determines $|b_0|^2$. We now will solve these equations. First, we consider the Green function of the s fermions:

$$G_s(\mathrm{i}\omega_n) = \frac{1}{\mathrm{i}\omega_n - E_\mathrm{f} - \lambda + \mathrm{i}|b_0|^2 \Delta_0 \operatorname{sgn} \omega_n}. \tag{4.1.30}$$

We obtain the retarded Green function by analytic continuation

$$G_s^\mathrm{R}(\omega) = G_s(\mathrm{i}\omega_n \to \omega + \mathrm{i}\delta) = \frac{1}{\omega - E_\mathrm{f} - \lambda + \mathrm{i}|b_0|^2 \Delta_0}, \tag{4.1.31}$$

and the spectral function $A_s(\omega)$,

$$A_s(\omega) = \frac{1}{\pi} \operatorname{Im} G_s^\mathrm{R}(\omega) = \frac{|b_0|^2 \Delta_0}{\pi \{(\omega - E_\mathrm{f} - \lambda)^2 + (|b_0|^2 \Delta_0)^2\}}, \tag{4.1.32}$$

behaves like a Lorentz curve.

With these results, we can write

$$G_s(\mathrm{i}\omega_n) = \int \mathrm{d}\omega \frac{A_s(\omega)}{\mathrm{i}\omega_n - \omega} \tag{4.1.33}$$

and the first term on the right-hand side of (4.1.28) becomes, using (4.1.32), (4.1.34)

$$\begin{aligned} \frac{1}{\beta} \sum_{\mathrm{i}\omega_n} G_s(\mathrm{i}\omega_n)\, \mathrm{e}^{\mathrm{i}\omega_n \delta} &= \int \mathrm{d}\omega\, A_s(\omega) \frac{1}{\beta} \sum_{\mathrm{i}\omega_n} \frac{\mathrm{e}^{\mathrm{i}\omega_n \delta}}{\mathrm{i}\omega_n - \omega} \\ &= \int \mathrm{d}\omega\, A_s(\omega) f(\omega). \end{aligned} \tag{4.1.34}$$

Performing the limit $T \to 0$, (4.1.28) becomes

$$\frac{1}{2} - \frac{1}{\pi} \tan^{-1}\left(\frac{E_\mathrm{f} + \lambda}{|b_0|^2 \Delta_0}\right) + |b_0|^2 = q. \tag{4.1.35}$$

Inserting $N = 2$ and $q = 1/2$, we obtain

$$|b_0|^2 = \frac{1}{\pi} \tan^{-1}\left(\frac{E_\mathrm{f} + \lambda}{|b_0|^2 \Delta_0}\right). \tag{4.1.36}$$

On the other hand, from (4.1.29) we obtain

$$\lambda = \frac{\Delta_0}{2\pi} \ln \frac{D^2}{(E_\mathrm{f} + \lambda)^2 + (|b_0|^2 \Delta_0)^2}. \tag{4.1.37}$$

The complete summation over ω_n would diverge, however, the summation is restricted to the region $|\omega_n| < D$ (D: cut-off of order of the band width of the conduction electrons) in the Green function (4.1.30).

Next, we consider the so-called Kondo limit. Then, E_f has a large negative value, the f electron (or s fermion) occupation is almost 1, and the probability

$|b_0|^2$ for a hole is small. This is the case for $|E_f + \lambda| \ll |b_0|^2 \Delta_0$ in (4.1.36), and expanding the $\tan^{-1}$, we obtain

$$|b_0|^4 \Delta_0 = \frac{E_f + \lambda}{\pi}. \tag{4.1.38}$$

Ignoring the term $(E_f + \lambda)^2$ in the logarithm in (4.1.37), we obtain

$$|b_0|^2 \Delta_0 = D\,\mathrm{e}^{-\pi\lambda/\Delta_0}. \tag{4.1.39}$$

Because $(E_f + \lambda)$ is small, we can set $\lambda = |E_f|$ in (4.1.39)

$$|b_0|^2 \Delta_0 = D\,\mathrm{e}^{-\pi|E_f|/\Delta_0}. \tag{4.1.40}$$

Inserting this into (4.1.38), the result is

$$\tilde{E}_f \equiv \lambda + E_f = \pi \frac{D^2}{\Delta_0} \mathrm{e}^{-2\pi|E_f|/\Delta_0}. \tag{4.1.41}$$

The Kondo limit corresponds to $|E_f| \gg \Delta_0$, and in this case the meaning of the term in the exponential is

$$2\pi \frac{|E_f|}{\Delta_0} = 2\pi \frac{|E_f|}{\pi \left\langle |V_k|^2 \right\rangle D(E_F)} = \frac{2|E_f|}{\left\langle |V_k|^2 \right\rangle D(E_F)} = \frac{2}{J_K D(E_F)}. \tag{4.1.42}$$

Here, J_K is the so-called Kondo coupling of the exchange interaction. It is the coupling constant between the conduction electron spins and the f-electron spin when in the intermediate state the f-electron state with energy $|E_f|$ becomes empty.

Using the same technique that has been applied when reducing the Hubbard model to the Heisenberg model, it is possible to obtain the s-d model (Kondo model) from the Anderson model (3.1.22):

$$H_{\mathrm{Kondo}} = \sum_{\boldsymbol{k},\sigma} (\varepsilon_{\boldsymbol{k}} - \mu) C^\dagger_{\boldsymbol{k}\sigma} C_{\boldsymbol{k}\sigma} + 2J_K \boldsymbol{s} \cdot \boldsymbol{S}. \tag{4.1.43}$$

Here,

$$\boldsymbol{s} = \frac{1}{2N_0} \sum_{\substack{\boldsymbol{k},\boldsymbol{k}' \\ \alpha,\beta}} C^\dagger_{\boldsymbol{k}\alpha} \boldsymbol{\sigma}_{\alpha\beta} C_{\boldsymbol{k}'\beta} \tag{4.1.44}$$

is the spin of the conduction electron occurring at the site of the impurity, and

$$\boldsymbol{S} = \frac{1}{2} \sum_{\alpha,\beta} f^\dagger_\alpha \boldsymbol{\sigma}_{\alpha\beta} f_\beta \tag{4.1.45}$$

is the f-spin. The reader might think at first glance that due to the constraint

$$\sum_{\alpha} f_{\alpha}^{\dagger} f_{\alpha} = 1 \tag{4.1.46}$$

the existence of a localized spin is guaranteed. However, owing to the anti-ferromagnetic-like influence of J_{K}, singlet states with the surrounding spins of the conduction electrons finally emerge. In the Anderson model the emergence of holes with finite probability $|b_0|^2$ describes the exchange interaction and, in turn, the singlet formation.

Considering the Green function of the f electron,

$$G_{\mathrm{f}}(\mathrm{i}\omega_n) = \frac{|b_0|^2}{\mathrm{i}\omega_n - \tilde{E}_{\mathrm{f}} + \mathrm{i}|b_0|^2 \Delta_0 \operatorname{sgn} \omega_n}, \tag{4.1.47}$$

we see that at the Fermi energy $\tilde{E}_{\mathrm{f}}$ a peak with width $|b_0|^2 \Delta_0$ and weight $|b_0|^2$ emerges. $|b_0|^2$ expresses the weight of the quasi-particle; usually, this factor is denoted by Z. The remaining $1 - Z$ is the incoherent background, so that at around $E = E_{\mathrm{f}}$ and $E_{\mathrm{f}} + U$, a broad peak emerges. The width $|b_0|^2 \Delta_0$ is the effective hybridization energy with the conduction electrons, and its inverse gives the characteristic time scale where the spin of the f-electron disappears due to the interplay with the conduction electrons. That is, we have indentified the Kondo temperature as

$$T_{\mathrm{K}} \cong |b_0|^2 \Delta_0 = D\, \mathrm{e}^{-\pi |E_{\mathrm{f}}|/\Delta_0} . \tag{4.1.48}$$

In the above discussion the coherent part of the f-electron is expressed as $f^{\dagger} = \langle b \rangle s^{\dagger}$ with Bose condensated b, and beside of this factor, the system is a free fermion model. Therefore, the system is (locally) a Fermi liquid. Indeed, because it has been confirmed that the ground state in the Kondo model is a singlet, and the low-energy properties can be described using the theory of a local Fermi liquid, we confirm that the above considerations are qualitatively correct. However, for the case when the slave bosons do not condense, locally, a non-Fermi liquid occurs. Is this possibility unrealistic?

The striking answer is 'no'. This possibility is indeed realized in the multi-channel Kondo problem. 'Multi-channel' means that the conduction electrons have a variety of channels (for example, of other degrees of freedom than spin, as orbitals etc.). On the other hand, we assume that the f-electron has spin S, and its orbital degrees of freedom freeze out. In this case, three different relations between S and the number of channels M become possible. Up to now, we have considered the case $S = 1/2$ and $M = 1$, where the localized spin and the spin of the conduction electrons form a singlet. In general for $2S = M$, singlet formation as shown in Fig. 4.1(a) emerges. However, for $2S > M$, the number of spins of conduction electrons is insufficient, and therefore $S - M/2$ localized spins remain (Fig. 4.1(b)). This is the so-called under-screening Kondo effect. In this case, the local Fermi liquid and the remaining local spins are decoupled asymptotically. Even more interesting is the case $2S < M$, the so-called over-screening Kondo effect (Fig. 4.1(c)). In this case,

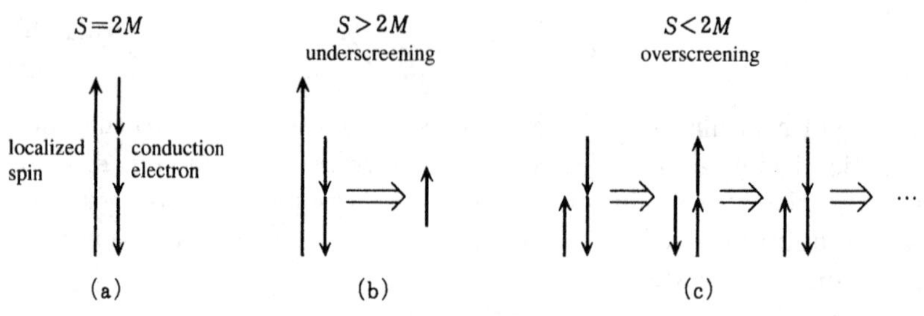

Fig. 4.1a–c. The multi-channel Kondo effect

because the conduction electrons screen the localized spins too strongely, localized spins again emerge. These are again screened by the conduction electrons, but due to the over-kill again local spins emerge, and so on. As a result, the ground state is degenerate, and the system becomes locally a non-Fermi liquid.

So, does there exist a saddle point solution describing this non-Fermi liquid state? Indeed, the saddle-point solution in a different $N \to \infty$ limit just corresponds to the picture described above. The multi-channel Kondo Hamiltonian is given by

$$H_{\text{Kondo}} = \sum_{m=1}^{M} \sum_{\sigma=\uparrow\downarrow} \sum_{\boldsymbol{k}} \varepsilon_{\boldsymbol{k}} C^{\dagger}_{\boldsymbol{k}m\sigma} C_{\boldsymbol{k}m\sigma} + \frac{2J_{\text{K}}}{N_0} \sum_{m=1}^{M} \sum_{\boldsymbol{k},\boldsymbol{k}'} \sum_{\alpha} C^{\dagger}_{\boldsymbol{k}m\alpha} \boldsymbol{\sigma}_{\alpha\beta} C_{\boldsymbol{k}'m\beta} \cdot \boldsymbol{S} , \tag{4.1.49}$$

with N_0 being the number of lattice points. M is the number of channels, and the spin $\boldsymbol{S}$ of the impurity is assumed to be 1/2.

Next, we imagine a fictitious Anderson Hamiltonian having the property that the effective theory deduced from it leads back to (4.1.49). This Hamiltonian is given by

$$H = \sum_{m} \sum_{\sigma} \sum_{\boldsymbol{k}} \varepsilon_{\boldsymbol{k}} C^{\dagger}_{\boldsymbol{k}m\sigma} C_{\boldsymbol{k}m\sigma} + E_{\text{f}} \sum_{\sigma} f^{\dagger}_{\sigma} f_{\sigma} + \frac{V}{\sqrt{N_0}} \sum_{\boldsymbol{k},m,\sigma} \left[f^{\dagger}_{\sigma} b_m C_{\boldsymbol{k}m\sigma} + \text{h.c.} \right] . \tag{4.1.50}$$

Here, f_σ and b_m are the fermions and bosons occurring in the slave-boson method, where the constraint in this case is given by

$$\sum_{m=1}^{M} b^{\dagger}_m b_m + \sum_{\sigma=1}^{2} f^{\dagger}_{\sigma} f_{\sigma} = 1 , \tag{4.1.51}$$

instead of (4.1.21). Here, the hole state is representated by M bosons, which is just a mathematical technique to derive the Kondo Hamiltonian (4.1.49). In this sense, the Anderson Hamiltonian (4.1.50) should be regarded as fictitious. Performing a perturbative expansion in V for the case when E_{f} is negative, by setting

$$J_{\mathrm{K}} = \frac{V^2}{|E_{\mathrm{f}}|} \tag{4.1.52}$$

the Kondo Hamiltonian (4.1.49) is regained.

We now re-calculate the path integral using the saddle-point method, where the sum in σ is now generalized and runs between $1, \ldots, N$, and simultaneously, we set $V \to (1/\sqrt{N})V$. That is, m runs from 1 to M, σ from 1 to N, and the limit $N \to \infty$ with finite $\gamma \equiv M/N$ is considered. Then, also after having performed the limit, the right-hand side of (4.1.51) is still 1, and not proportional to N as was assumed in (4.1.23). Consequently, $b_m^\dagger b_m$ becomes of order $1/N$, and in the limit $N \to \infty$, Bose condensation does not occur. Writing Q for the left-hand side of (4.1.51), the constraint can be expressed by

$$Z = \lim_{\lambda \to \infty} \mathrm{e}^{\beta\lambda}\, \mathrm{Tr}[\mathrm{e}^{-\beta H(\lambda)} Q]\,, \tag{4.1.53}$$

with

$$H(\lambda) = H + \lambda Q\,. \tag{4.1.54}$$

In terms of the path integral, this expression reads

$$Z = \lim_{\lambda \to \infty} \mathrm{e}^{\beta\lambda} \int \mathcal{D}C^\dagger \mathcal{D}C \mathcal{D}f^\dagger \mathcal{D}f \mathcal{D}b^\dagger \mathcal{D}b \exp\left(- \int_0^\beta L\, \mathrm{d}t \right)\,, \tag{4.1.55}$$

$$\begin{aligned} L = &\sum_{m,\sigma,\boldsymbol{k}} C_{\boldsymbol{k}m\sigma}^\dagger (\partial_\tau + \varepsilon_{\boldsymbol{k}} - \mu) C_{\boldsymbol{k}m\sigma} + \sum_\sigma f_\sigma^\dagger (\partial_\tau + \lambda + E_{\mathrm{f}}) f_\sigma \\ &+ \sum_m b_m^\dagger (\partial_\tau + \lambda) b_m + \frac{V}{\sqrt{N N_0}} \sum_{\boldsymbol{k},m,\sigma} \left[f_\sigma^\dagger b_m C_{\boldsymbol{k}m\sigma} + \mathrm{h.c.} \right]. \end{aligned} \tag{4.1.56}$$

We now integrate out $C^\dagger$ and C. It is sufficient to complete the square for the Grassmann numbers $C^\dagger$ and C. After Fourier transformation, writing $\xi_{\boldsymbol{k}} = \varepsilon_{\boldsymbol{k}} - \mu$, the linear transformations

$$\begin{aligned} \tilde{C}_{\boldsymbol{k}m\sigma}(\mathrm{i}\omega_n) &= C_{\boldsymbol{k}m\sigma}(\mathrm{i}\omega_n) + \frac{1}{-\mathrm{i}\omega_n + \xi_{\boldsymbol{k}}} \frac{V}{\sqrt{N_0 N}} \left(b_m^\dagger f_\sigma \right)_{\mathrm{i}\omega_n} \\ \tilde{C}_{\boldsymbol{k}m\sigma}^\dagger(\mathrm{i}\omega_n) &= C_{\boldsymbol{k}m\sigma}^\dagger(\mathrm{i}\omega n) + \frac{1}{-\mathrm{i}\omega_n + \xi_{\boldsymbol{k}}} \frac{V}{\sqrt{N_0 N}} \left(f_\sigma^\dagger b_m \right)_{\mathrm{i}\omega_n} \end{aligned} \tag{4.1.57}$$

of the integration variables are appropriate, and when the integral is performed, we are left with

$$\frac{V^2}{N} \sum_{m,\sigma} \sum_{\mathrm{i}\omega_n} \left(f_\sigma^\dagger b_m \right)_{\mathrm{i}\omega_n} G(\mathrm{i}\omega_n) \left(b_m^\dagger f_\sigma \right)_{\mathrm{i}\omega_n}\,. \tag{4.1.58}$$

The Green function is given by

$$G(\mathrm{i}\omega_n) = \frac{1}{N_0} \sum_{\boldsymbol{k}} \frac{1}{\mathrm{i}\omega_n - \xi_{\boldsymbol{k}}} . \tag{4.1.59}$$

Expressed as an imaginary time integration, this becomes

$$\frac{V^2}{N} \sum_{m,\sigma} \int \mathrm{d}\tau\, \mathrm{d}\tau'\, f_\sigma^\dagger(\tau) b_m(\tau) G(\tau - \tau') b_m^\dagger(\tau') f_\sigma(\tau') . \tag{4.1.60}$$

The above calculation corresponds to the following physical picture. For the localized spins, the conduction electrons are a heat bath. (4.1.60) describes the process when an electron appears in the heat bath at time τ', and after having wandered around in the heat bath during the time $\tau - \tau'$, it reappears at the impurity site at time τ. The heat bath causes a non-local interaction in the time direction, that is, dissipation and irreversibility.

Next, we analyse (4.1.60) further using two kinds of Stratonovich–Hubbard fields, $\Phi_{f,\sigma}$ and $\Phi_{b,m}$. The complete action integral can then be written as

$$\begin{aligned} A = &\int \mathrm{d}\tau \left[\sum_\sigma f_\sigma^\dagger (\partial_\tau + E_\mathrm{f} + \lambda) f_\sigma + \sum_m b_m^\dagger (\partial_\tau + \lambda) b_m \right] \\ &- \frac{V^2}{N} \int \mathrm{d}\tau\, \mathrm{d}\tau'\, G(\tau - \tau') \sum_{m,\sigma} \left[\Phi_{bm}(\tau, \tau') f_\sigma^\dagger(\tau) f_\sigma(\tau') \right. \\ &\left. - \Phi_{f\sigma}(\tau', \tau) b_m^\dagger(\tau') b_m(\tau) - \Phi_{f\sigma}(\tau', \tau) \Phi_{bm}(\tau, \tau') \right] . \end{aligned} \tag{4.1.61}$$

In the limit $M, N \to \infty$, the whole action is of $\mathcal{O}(N)$, and therefore the saddle-point method can be applied.

Assuming that the saddle-point solution is symmetric for the translation in time, we can write $\Phi_{f,\sigma} = \Phi_f(\tau - \tau')$ and $\Phi_{b,m} = \Phi_b(\tau - \tau')$. The action then becomes

$$\begin{aligned} A = &\sum_\sigma \sum_{\mathrm{i}\omega_n} f_\sigma^\dagger(\mathrm{i}\omega_n) \left[-\mathrm{i}\omega_n + E_\mathrm{f} + \lambda + \Sigma_f(\mathrm{i}\omega_n) \right] f_\sigma(\mathrm{i}\omega_n) \\ &+ \sum_m \sum_{\mathrm{i}\omega_l} b_m^\dagger(\mathrm{i}\omega_l) \left[-\mathrm{i}\omega_l + \lambda + \Sigma_b(\mathrm{i}\omega_l) \right] b_m(\mathrm{i}\omega_l) \\ &+ V^2 M \beta \int \mathrm{d}\tau\, \Phi_f(\tau) \Phi_b(-\tau) . \end{aligned} \tag{4.1.62}$$

Here, we used the two definitions

$$\Sigma_f(\tau) = -\gamma V^2 G(\tau) \Phi_b(\tau)$$

$$\Sigma_b(\tau) = V^2 G(\tau) \Phi_f(-\tau) \tag{4.1.63}$$

and their Fourier transformations are defined as $\sum(\mathrm{i}\omega) = \int_0^\beta \mathrm{d}\tau\, \mathrm{e}^{\mathrm{i}\omega\tau} \sum(\tau)$. $\Phi_b(\tau)$ and $\Phi_f(\tau)$ are determined by setting $\delta A = 0$, and going back to (4.1.61), they are given by

$$\Phi_b(\tau,\tau') = -\left\langle T_\tau b_m(\tau) b_m^\dagger(\tau')\right\rangle$$

$$\Phi_f(\tau,\tau') = -\left\langle T_\tau f_\sigma(\tau) f_\sigma^\dagger(\tau')\right\rangle . \tag{4.1.64}$$

These are the bosonic and the fermionic Green functions, respectively.

We obtain

$$\Phi_b(\mathrm{i}\omega_l) = \left[\mathrm{i}\omega_l - \lambda - \Sigma_b(\mathrm{i}\omega_l)\right]^{-1}$$

$$\Phi_f(\mathrm{i}\omega_n) = \left[\mathrm{i}\omega_n - E_f - \lambda - \Sigma_f(\mathrm{i}\omega_n)\right]^{-1} , \tag{4.1.65}$$

on the other hand, performing a Fourier transformation in (4.1.63), we obtain for example for $\sum_f$

$$\begin{aligned}
\Sigma_f(\mathrm{i}\omega_n) &= -\gamma V^2 \int_0^\beta \mathrm{d}\tau\, \mathrm{e}^{\mathrm{i}\omega_n \tau} G(\tau)\Phi_b(\tau) \\
&= -\frac{1}{\beta^2}\gamma V^2 \sum_{\mathrm{i}\omega_l, \mathrm{i}\omega_m} \int_0^\beta \mathrm{d}\tau\, \mathrm{e}^{(\mathrm{i}\omega_n - \mathrm{i}\omega_m - \mathrm{i}\omega_l)\tau} G(\mathrm{i}\omega_m)\Phi_b(\mathrm{i}\omega_l) \\
&= -\gamma V^2 \frac{1}{\beta}\sum_{\mathrm{i}\omega_m} G(\mathrm{i}\omega_m)\Phi_b(\mathrm{i}\omega_n - \mathrm{i}\omega_m) .
\end{aligned} \tag{4.1.66}$$

Inserting the spectral decomposition of Φ_b

$$\Phi_b(\mathrm{i}\omega_n - \mathrm{i}\omega_m) = \int \mathrm{d}\omega \frac{A_b(\omega)}{\mathrm{i}\omega_n - \mathrm{i}\omega_m - \omega} \tag{4.1.67}$$

and performing the summation in ω_n, (4.1.66) becomes

$$\Sigma_f(\mathrm{i}\omega_n) = \frac{\gamma V^2}{N_0}\sum_{\boldsymbol{k}} \int \mathrm{d}\omega\, A_b(\omega) \frac{f(-\xi_{\boldsymbol{k}}) + n(\omega)}{\mathrm{i}\omega_n - \xi_{\boldsymbol{k}} - \omega} . \tag{4.1.68}$$

Here, we used (4.1.59).

We now consider the spectral function of the bosons and the fermions further. Because both are dominated by the region $\omega \simeq \lambda$, we define

$$\bar{A}_{f,b}(\omega) = A_{f,b}(\omega + \lambda) . \tag{4.1.69}$$

Here, the limit $\lambda \to \infty$ can easily be performed. Then, the thermal Green functions with $\tau > 0$ can be expressed as

$$\Phi_f(\tau) = \int_{-\infty}^{\infty} \mathrm{d}\omega\, \mathrm{e}^{-(\omega+\lambda)\tau}\left[1 - f(\omega+\lambda)\right]\bar{A}_f(\omega)$$

$$\Phi_b(\tau) = \int_{-\infty}^{\infty} \mathrm{d}\nu\, \mathrm{e}^{-(\nu+\lambda)\tau}\left[1 + n(\nu+\lambda)\right]\bar{A}_b(\nu) \tag{4.1.70}$$

respectively. Here we used the fact that for $\tau > 0$

$$\frac{1}{\beta} \sum_{\substack{\mathrm{i}\omega_n \\ \text{fermion}}} \frac{\mathrm{e}^{--\mathrm{i}\omega_n \tau}}{\mathrm{i}\omega_n - x} = 1 - f(x) \tag{4.1.71}$$

and

$$\frac{1}{\beta} \sum_{\substack{\mathrm{i}\nu_n \\ \text{boson}}} \frac{\mathrm{e}^{-\mathrm{i}\nu_n \tau}}{\mathrm{i}\nu_n - x} = 1 + n(x) \tag{4.1.72}$$

are valid.

Then, $\bar{A}_f(\omega)$ and $\bar{A}_b(\nu)$ are

$$\begin{aligned}
\bar{A}_f(\omega) &= Z(\lambda)^{-1} \sum_{i,j} \left| \langle j; Q_j | f_\sigma^\dagger | i; Q_i \rangle \right|^2 \left(\mathrm{e}^{-\beta(E_j + \lambda Q_j)} + \mathrm{e}^{-\beta(E_i + \lambda Q_i)} \right) \\
&\quad \times \delta(\omega - (E_j - E_i)) , \\
\bar{A}_b(\nu) &= Z(\lambda)^{-1} \sum_{i,j} \left| \langle j; Q_j \left| b_m^\dagger \right| i; Q_i \rangle \right|^2 \left(\mathrm{e}^{-\beta(E_j + \lambda Q_j)} + \mathrm{e}^{-\beta(E_i + \lambda Q_i)} \right) \\
&\quad \times \delta(\nu - (E_j - E_i)) ,
\end{aligned} \tag{4.1.73}$$

respectively.

Q_i and Q_j are the values of Q in the states i and j, fulfilling the relation $Q_j = Q_i + 1$. $Z(\lambda)$ is defined as

$$Z(\lambda) = \mathrm{Tr}\left[\mathrm{e}^{-\beta H(\lambda)} \right] = \sum_{Q=0}^{\infty} \mathrm{e}^{-\beta \lambda Q} Z(Q) . \tag{4.1.74}$$

Performing the limit $\lambda \to \infty$, only the contributions $Q_i = 0$ and $Q_j = 1$ remain, and (4.1.74) becomes

$$\begin{aligned}
\bar{A}_f(\omega) &= Z(Q=0)^{-1} \sum_{i,j} \left| \langle j; Q_j = 1 \left| f_\sigma^\dagger \right| i; Q_i = 0 \rangle \right|^2 \mathrm{e}^{-\beta E_i} \delta(\omega - (E_j - E_i)) , \\
\bar{A}_b(\omega) &= Z(Q=0)^{-1} \sum_{i,j} \left| \langle j; Q_j = 1 \left| b_m^\dagger \right| i; Q_i = 0 \rangle \right|^2 \mathrm{e}^{-\beta E_i} \delta(\omega - (E_j - E_i)) .
\end{aligned} \tag{4.1.75}$$

Here, $Z(Q = 0)$ is the partition function of the conduction electrons without impurities. Furthermore, performing the limit $T \to 0$, only the ground state $Q_i = 0$ contributes the sum in i, and therefore with $Z(Q = 0)^{-1} \simeq \mathrm{e}^{\beta E_0}$ we obtain

$$\bar{A}_f(\omega) = \sum_j \left|\langle j; Q = 1 \left| f_\sigma^\dagger \right| 0; Q = 0 \rangle\right|^2 \delta(\omega - E_j) \,,$$
$$\bar{A}_b(\omega) = \sum_j \left|\langle j; Q = 1 \left| b_m^\dagger \right| 0; Q = 0 \rangle\right|^2 \delta(\omega - E_j) \,. \tag{4.1.76}$$

Calling $E_{\rm G}$ the ground state energy in the subspace with $Q = 1$, we conclude that for $T \to 0$, only in the case $\omega \geq E_{\rm G}$ are $\bar{A}_f(\omega)$ and $\bar{A}_b(\omega)$ finite. As we have seen above, $\bar{A}_{b,f}(\omega)$ converges to a finite function, and therefore, due to $\omega \simeq \lambda$, $n(\omega)$ appearing in the ω integral (4.1.68), can be ignored. Having this in mind, we measure the energy ω from λ.

Using this, (4.1.68) can be written as

$$\begin{aligned}\Sigma_f(\mathrm{i}\omega_n) &= \gamma V^2 \int \mathrm{d}\xi \, D(\xi) f(\xi) \int \frac{\bar{A}_b(\omega) \, \mathrm{d}\omega}{\mathrm{i}\omega_n + \xi - \omega} \\ &= \gamma V^2 \int \mathrm{d}\xi \, D(\xi) f(\xi) \Phi_b(\mathrm{i}\omega_n + \xi) \,.\end{aligned} \tag{4.1.77}$$

Here, we introduced the density of states $D(\xi) = 1/N_0 \sum_{\boldsymbol{k}} \delta(\xi - \xi_{\boldsymbol{k}})$. Performing analytical continuation and setting $\mathrm{i}\omega_n \to \omega + \mathrm{i}\delta$, the relation

$$\Sigma_f^{\rm R}(\omega) = \frac{\gamma\Gamma}{\pi} \int \mathrm{d}\xi \, f(\xi) \Phi_b^{\rm R}(\omega + \xi) \tag{4.1.78}$$

between the retarded Green functions is obtained. Notice that by shifting the origin of ω about λ, in equation (4.1.65), λ can be set to zero. The density of states $D(\xi)$ is set constant $(= D)$, and $\Gamma = \pi D V^2$ has been introduced, which just equals Δ_0 in (4.1.4). The same considerations can be performed for the self-energy Σ_b of the boson, with the result

$$\Sigma_b^{\rm R}(\omega) = \frac{\Gamma}{\pi} \int \mathrm{d}\xi \, f(\xi) \Phi_f^{\rm R}(\xi) \,. \tag{4.1.79}$$

We now derived with (4.1.65), (4.1.78) and (4.1.79) the equations that have to be solved. The approximation that has been performed corresponds to the non-crossing approximation (NCA). The solution of these equations can be derived as follows. We will perform the calculation at zero temperature $T = 0$. First, we differentiate (4.1.78) and (4.1.79) with respect to ω

$$\begin{aligned}\frac{\partial \Sigma_f^{\rm R}(\omega)}{\partial \omega} &= \frac{\gamma\Gamma}{\pi} \int \mathrm{d}\xi \frac{\partial f(\xi - \omega)}{\partial w} \Phi_b^{\rm R}(\xi) \\ &= \frac{\gamma\Gamma}{\pi} \int \mathrm{d}\xi \, \delta(\xi - \omega) \Phi_b^{\rm R}(\xi) = \frac{\gamma\Gamma}{\pi} \Phi_b^{\rm R}(\omega)\end{aligned} \tag{4.1.80}$$

$$\frac{\partial \Sigma_b^{\rm R}(\omega)}{\partial \omega} = \frac{\Gamma}{\pi} \Phi_f^{\rm R}(\omega) \,. \tag{4.1.81}$$

Now, we impose the initial conditions that Σ_f^{R} and Σ_b^{R} vanish at the lower-bound cut-off $\omega = -D$. Defining

$$\begin{aligned} Y_f(\omega) &= -\left[\Phi_f^{\mathrm{R}}(\omega)\right]^{-1} = -\omega + E_{\mathrm{f}} + \Sigma_f^{\mathrm{R}}(\omega)\,, \\ Y_b(\omega) &= -\left[\Phi_b^{\mathrm{R}}(\omega)\right]^{-1} = -\omega + \Sigma_b^{\mathrm{R}}(\omega)\,, \end{aligned} \tag{4.1.82}$$

then equations (4.1.80) and (4.1.81) can be expressed as

$$\begin{aligned} \frac{\partial Y_f(\omega)}{\partial \omega} &= -1 + \frac{\partial \Sigma_f^{\mathrm{R}}(\omega)}{\partial \omega} = -1 - \frac{\gamma \Gamma}{\pi} Y_b^{-1}(\omega)\,, \\ \frac{\partial Y_b(\omega)}{\partial \omega} &= -1 + \frac{\partial \Sigma_b^{\mathrm{R}}(\omega)}{\partial \omega} = -1 - \frac{\Gamma}{\pi} Y_f^{-1}(\omega)\,. \end{aligned} \tag{4.1.83}$$

The ratio of the two equations (4.1.83) becomes

$$\frac{\mathrm{d}Y_f}{\mathrm{d}Y_b} = \frac{1 + \dfrac{\gamma \Gamma}{\pi} Y_b^{-1}}{1 + \dfrac{\Gamma}{\pi} Y_f^{-1}} \tag{4.1.84}$$

and, after integration, we obtain

$$Y_f + \frac{\Gamma}{\pi} \ln \frac{Y_f}{D} + C = Y_b + \frac{\gamma \Gamma}{\pi} \ln \frac{Y_b}{D}\,. \tag{4.1.85}$$

C is an integration constant, and using the initial conditions $Y_f(-D) = D + E_{\mathrm{f}}$ and $Y_b(-D) = D$ at $\omega = -D$, this constant is fixed to be

$$C = -E_{\mathrm{f}} - \frac{\Gamma}{\pi} \ln \left(1 + \frac{E\mathrm{f}}{D}\right)\,. \tag{4.1.86}$$

With regard to $E_{\mathrm{f}}/D \ll 1$, we obtain $C \simeq -E_{\mathrm{f}}$.

We now assume that $\bar{A}_f(\omega)$ and $\bar{A}_b(\omega)$ are diverging at $\omega \to E_{\mathrm{G}}$ following a power-law behaviour. As can be seen in (4.1.73), these quantities describe the response of the conduction electrons when a fermion or a boson is created, where a singularity similar to that at the X-ray absorption edge is expected. This will be self-consistently justified a posteriori. Here, we will proceed using this as an assumption.

Then, owing to the relation $\bar{A}_{f,b}(\omega) = -(1/\pi)\,\mathrm{Im}\,\Phi_{f,b}^{\mathrm{R}}$, we can conclude $[\Phi_{f,b}^{\mathrm{R}}(\omega)]^{-1} \to 0$ for $\omega \to E_{\mathrm{G}}$. Furthermore, because for $\omega < E_{\mathrm{G}}$, $\bar{A}_{f,b}(\omega)$ vanishes, in this region, $\Phi_{f,b}^{\mathrm{R}}$ and $Y_{f,b}(\omega)$ are real. We will integrate (4.1.83) under these conditions. (4.1.85) can be interpreted as an equation expressing Y_f as a function of Y_b. Thinking of Y_f as $Y_f = Y_f(Y_b)$, for the second equation in (4.1.83) after integration with respect to Y_b we obtain

$$\int_{-D}^{\omega} \mathrm{d}\varepsilon = -\int_{D}^{Y_b} \frac{\mathrm{d}x}{1 + \Gamma/\pi Y_f^{-1}(x)} = \frac{\Gamma}{\pi} \int_{0}^{Y_b} \frac{\mathrm{d}x}{\Gamma/\pi + Y_f(x)} - (Y_b - D)\,. \tag{4.1.87}$$

This equals

$$\omega = -Y_b - \frac{\Gamma}{\pi} \int_{Y_b}^{D} \frac{\mathrm{d}x}{\Gamma/\pi + Y_f(x)} . \tag{4.1.88}$$

Because $Y_b = 0$ for $\omega = E_{\mathrm{G}}$, from (4.1.88) we obtain

$$E_{\mathrm{G}} = -\frac{\Gamma}{\pi} \int_{0}^{d} \frac{\mathrm{d}x}{\Gamma/\pi + Y_f(x)} . \tag{4.1.89}$$

Subtracting (4.1.89) from (4.1.88), we obtain

$$\omega - E_{\mathrm{G}} = -\int_{0}^{Y_b} \mathrm{d}x \frac{Y_f(x)}{\Gamma/\pi + Y_f(x)} . \tag{4.1.90}$$

Now, because for small $|\omega - E_{\mathrm{G}}|$, $Y_f(\omega)$ as well as $Y_b(\omega)$ is small, we conclude that the logarithmic term in (4.1.85) is dominant, and obtain

$$\frac{Y_f(\omega)}{D} \mathrm{e}^{-\pi E_{\mathrm{f}}/\Gamma} = \left[\frac{Y_b(\omega)}{D}\right]^{\gamma} . \tag{4.1.91}$$

Defining the Kondo temperature T_{NCA} in the framework of the NCA approximation as

$$T_{\mathrm{NCA}} = D(\gamma\Gamma/\pi D)^{\gamma} \, \mathrm{e}^{\pi E_{\mathrm{f}}/\Gamma} \tag{4.1.92}$$

(4.1.91) can then be written as

$$\frac{Y_f(\omega)}{T_{\mathrm{NCA}}} = \left[\frac{\pi Y_b(\omega)}{\gamma\Gamma}\right]^{\gamma} . \tag{4.1.93}$$

Inserting this equation into (4.1.90) and ignoring $Y_f(x)$ compared with Γ/π in the denominator, we obtain

$$\omega - E_{\mathrm{G}} \cong -\frac{\pi}{\Gamma} T_{\mathrm{NCA}} \frac{\gamma\Gamma}{\pi(\gamma+1)} \left[\frac{\pi Y_b(\omega)}{\gamma\Gamma}\right]^{\gamma+1} . \tag{4.1.94}$$

We conclude that in the limit $E_{\mathrm{G}} - \omega \to 0$, we obtain

$$\begin{aligned} Y_b(\omega) &\sim |\omega - E_{\mathrm{G}}|^{1/(\gamma+1)} , \\ Y_f(\omega) &\sim [Y_b(\omega)]^{\gamma} \sim |\omega - E_{\mathrm{G}}|^{\gamma/(\gamma+1)} , \end{aligned} \tag{4.1.95}$$

and from these results, we obtain

$$\begin{aligned} \Phi_b^{\mathrm{R}}(\omega) &\sim |\omega - E_{\mathrm{G}}|^{-1/(\gamma+1)} , \\ \Phi_f^{\mathrm{R}}(\omega) &\sim |\omega - E_{\mathrm{G}}|^{-\gamma/(\gamma+1)} . \end{aligned} \tag{4.1.96}$$

In such a way, the Green function of the fermions and bosons develops a singularity at E_{G} that is characterized by the exponent $\gamma = M/N$.

There still remains the work to calculate the specific heat, susceptibility and the resistance at low temperature using the Green function that we just derived. Therefore, the considerations that we made so far are only the starting point for the whole story. However, as can easily be imagined, the power-law behaviour at the singularity will cause also the power-law behaviour as the temperature dependence of these physical quantities. Furthermore, it has been shown that this models reproduced the properties of the non-Fermi liquid fixed point of the $M > 2S$ over-screening case. For more details, the reader is referred to [13]. We conclude that for the multi-channel Kondo effect, the saddle-point method can also be applied. However, we have to bear in mind that the choice of the saddle-point solution must be done using physical arguments and by combining it with other theoretical methods.

4.2 Dynamical Mean Field Theory

In the previous section we considered the problem of local electronic correlation at the location of an impurity. In this section, this point of view will be applied to the Hubbard model, that is, the problem of electron correlation at all sites. The principal ideas of this method already appear in the discussion of the Ising model using mean field theory.

With $\sigma_i = \pm 1$, we consider the following Hamiltonian:

$$H = -\sum_{i,j} J_{ij}\sigma_i\sigma_j - h\sum_i \sigma_i \,. \tag{4.2.1}$$

We set $J_{ij} > 0$. Now, we concentrate on one spin σ_i, and notice that the behaviour of this σ_i is triggered by the other spins via the coupling J_{ij}. We call H_i all the terms directly related to σ_i:

$$H_i = -\left(2\sum_j J_{ij}\sigma_j + h\right)\sigma_i \equiv -h_i^{\text{eff}}(\{\sigma_j\})\sigma_i \,. \tag{4.2.2}$$

Here, $h^{\text{eff}}(\{\sigma_i\})$ is the effective magnetic field depending on all spins except σ_i. At finite temperature T, this is the mean field approximation in the

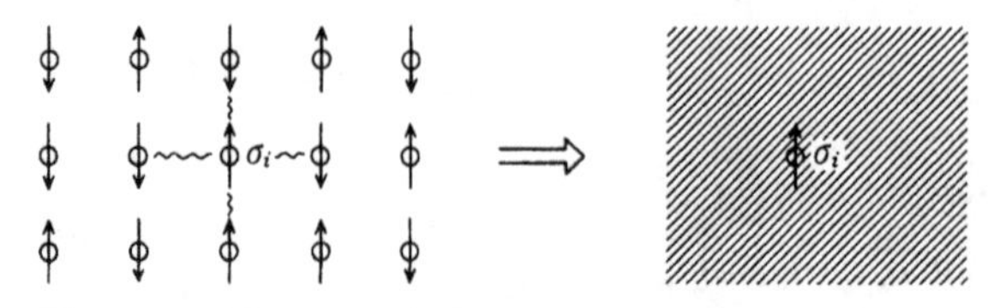

Fig. 4.2. The idea of the mean field theory

thermal equilibrium. Intuitively, as shown in Fig. 4.2, the individuality of the spins is washed out and replaced by a uniform medium:

$$\langle h_i^{\text{eff}}(\{\sigma_j\})\rangle = 2\sum_j J_{ij}\langle\sigma_j\rangle + h\,. \tag{4.2.3}$$

$\langle\sigma_j\rangle$ can be assumed to be independent of j, as well as $\langle h_i^{\text{eff}}\rangle$:

$$\langle h^{\text{eff}}\rangle = 2\Big(\sum_j J_{ij}\Big)\langle\sigma\rangle + h \equiv 2\bar{J}\langle\sigma\rangle + h\,. \tag{4.2.4}$$

However, because $\langle\sigma\rangle$ corresponds to the expectation value $\langle\sigma_i\rangle$, we obtain the following self-consistent equation for $\langle\sigma\rangle$:

$$\langle\sigma\rangle = \langle\sigma_i\rangle = \tanh\frac{\langle h_{\text{eff}}\rangle}{T} = \tanh\left[\frac{2\bar{J}\langle\sigma\rangle + h}{T}\right]. \tag{4.2.5}$$

Setting the external field h to zero, we obtain from (4.2.5) the well-known result that for $T < T_c = 2\bar{J}$, spontaneous magnetization $\langle\sigma\rangle \neq 0$ emerges. For $T > T_c$, the influence of $\bar{J}$ becomes visible in the susceptibility $\chi(T)$. Under an infinitesimal external magnetic field h, also $\langle\sigma\rangle = \chi(T)h$ is small, and therefore when the tanh in (4.2.5) is expanded

$$\langle\sigma\rangle \simeq \frac{2\bar{J}\langle\sigma\rangle + h}{T} \tag{4.2.6}$$

we obtain Curie's law:

$$\chi(T) = \frac{1}{T - T_c}\,. \tag{4.2.7}$$

Now, under what conditions is the mean field approximation justified? As can be seen in equation (4.2.2), h_i^{eff} is the sum of all spins σ_i connected to σ_j via the coupling J_{ij}. Therefore, when the number M of the spins σ_i becomes large, following the central limit theorem, the fluctuations around the expectation value become smaller by the order $M^{-1/2}$. For example, supposing for J_{ij} only a nearest-neighbour interaction, then in d dimensions M is given by $M = 2d$, and $M \to \infty$ corresponds to $d \to \infty$.

We now generalize the above considerations to the problem of electronic correlations. The Ising model is a classical model, and the spins σ_i have no quantum dynamics. Therefore, $\langle h^{\text{eff}}\rangle$ takes a constant value. On the other hand, electrons are quantum mechanical particles, and therefore also the 'effective magnetic field' is a dynamical field. In the framework of path integrals, this corresponds to (imaginary) time dependence.

We again consider the Hubbard model

$$H = -\sum_{i,j,\sigma} t_{ij} C_{i\sigma}^{\dagger} C_{j\sigma} + U\sum_i n_{i\uparrow}n_{i\downarrow}\,. \tag{4.2.8}$$

The analogy to (4.2.1) is given by the correspondence $J_{ij} \leftrightarrow t_{ij}$. However, different from the Ising model, it is impossible to consider $\langle C^{\dagger}_{i\sigma}\rangle$ or $\langle C_{i\sigma}\rangle$. This is due to the fact that the mean value of a fermionic operator is zero.

When we focus on one electron at site i, the effect of t_{ij} shows up in the following way. An electron that travels at time τ' from site i to site j is wandering around the time $\tau - \tau'$ (during this time, it feels the interaction with the other electrons) before returning again to i at time τ. This effect is represented by the propagator (Green function) $\mathcal{G}_0(\tau - \tau')$, and the effective action for the electron at site i is given by

$$
\begin{aligned}
A_{\text{eff}} = &- \int_0^{\beta} \mathrm{d}\tau \int_0^{\beta} \mathrm{d}\tau' \sum_{\sigma} C^{\dagger}_{i\sigma}(\tau)\mathcal{G}_0^{-1}(\tau - \tau')C_{i\sigma}(\tau') \\
&+ U \int_0^{\beta} \mathrm{d}\tau\, n_{i\uparrow}(\tau) n_{i\downarrow}(\tau) .
\end{aligned}
\tag{4.2.9}
$$

So, what might be the equation that self-consistently describes $\mathcal{G}_0(\tau - \tau')$? The starting point for the derivation is the self-energy $\sum_{ij}(i\omega_n)$ of the electron. $\sum_{ij}(i\omega_n)$ arises due to the Coulomb interaction U between the electrons. The idea of the mean field approximation is to implement its effect in a single-site model as given in (4.2.9). Similar to the Ising model, the limit where this approximation becomes exact is $d \to \infty$. In this limit, from the requirement that $\mathcal{G}_0^{-1} \propto \sum_j t_{ij}^2$ remains finite, it follows that t_{ij} must scale as $t_{ij} \propto d^{-1/2}$. Then, the self energy $\sum_{ij}(i \neq j)$ between different sites becomes at least $d^{-1/2}$ smaller compared with $\sum_{ii}$. That is, in the limit $d \to \infty$, the self-energy becomes $\sum_{ij}(\omega) = \delta_{ij} \sum(\omega)$.

This corresponds to $\boldsymbol{k}$-independent self-energy, and the Green function of the electrons becomes

$$
G(\boldsymbol{k}, \mathrm{i}\omega_n) = \frac{1}{\mathrm{i}\omega_n - \varepsilon_{\boldsymbol{k}} + \mu - \Sigma(\mathrm{i}\omega_n)} . \tag{4.2.10}
$$

The local Green function at site i can be expressed as

$$
G_{ii}(\mathrm{i}\omega_n) = \frac{1}{V}\sum_{\boldsymbol{k}} G(\boldsymbol{k}, \mathrm{i}\omega_n) = \int \mathrm{d}\varepsilon \frac{D(\varepsilon)}{\mathrm{i}\omega_n + \mu - \Sigma(\mathrm{i}\omega_n) - \varepsilon} , \tag{4.2.11}
$$

where the density of states per spin and volume is given by $D(\varepsilon) = 1/V \sum_{\boldsymbol{k}} \delta(\varepsilon - \varepsilon_{\boldsymbol{k}})$.

Defining the function $F(z)$ of the imaginary variable z by

$$
F(z) = \int \mathrm{d}\varepsilon \frac{D(\varepsilon)}{z - \varepsilon} , \tag{4.2.12}
$$

we obtain $G_{ii}(\mathrm{i}\omega_n) = F(\mathrm{i}\omega_n + \mu - \sum(\mathrm{i}\omega_n))$, and the inverse function of $F(z)$ is given by

$$
\mathrm{i}\omega_n + \mu - \Sigma(\mathrm{i}\omega_n) = F^{-1}(G_{ii}(\mathrm{i}\omega_n)) . \tag{4.2.13}
$$

The self-consistent equation is obtained by requiring that the Green function $G(\mathrm{i}\omega_n)$ obtained from (4.2.9) agrees with $G_{ii}(\mathrm{i}\omega_n)$. That is, inserting the self-energy obtained in (4.2.10) $\sum(\mathrm{i}\omega_n) = \mathcal{G}_0^{-1}(\mathrm{i}\omega_n) - G^{-1}(\mathrm{i}\omega_n)$ into (4.2.13), we obtain

$$\mathcal{G}_0^{-1}(\mathrm{i}\omega_n) = \mathrm{i}\omega_n + \mu + G^{-1}(\mathrm{i}\omega_n) - F^{-1}(G(\mathrm{i}\omega_n))\,. \tag{4.2.14}$$

The self-consistency condition is the following. The Green function $G(\mathrm{i}\omega_n)$ calculated in the local model (4.2.9), inserted into the right-hand side of (4.2.14), must agree with the Green function $\mathcal{G}_0^{-1}(\mathrm{i}\omega_n)$ that has been postulated in (4.2.9).

Many different types of state densites $D(\varepsilon)$ can be assumed; some representative examples are a a half circle

$$D(\varepsilon) = \frac{1}{2\pi t^2}\sqrt{4t^2 - \varepsilon^2}$$

or the Lorentzian density of states

$$D(\varepsilon) = \frac{t}{\pi(t^2 + \varepsilon^2)}\,.$$

Especially for the Lorentzian density of states, because $F(z)$ becomes $F(z) = (z + \mathrm{i}t\,\mathrm{sgn}\,\mathrm{Im}\,z)^{-1}$, (4.2.14) reads

$$\mathcal{G}_0^{-1}(\mathrm{i}\omega_n) = \mathrm{i}\omega_n + \mu + \mathrm{i}t\,\mathrm{sgn}\,\omega_n\,. \tag{4.2.15}$$

The right-hand side does not contain $G(\mathrm{i}\omega_n)$. Therefore, it is not necessary to solve the self-consistency equation, and it is sufficient to insert (4.2.15) into (4.2.9) and to solve the one-impurity problem. However, the disadvantage of the Lorentzian density of states is that the moments of ε diverge, corresponding to the limit of a large effective bandwidth. Therefore, for example the interesting problem of the Mott–Hubbard transition cannot be discussed using this density.

In practice, the most difficult step when a dynamical mean field theory is constructed is to calculate the impurity Green function $G(\mathrm{i}\omega_n)$ of the impurity problem (4.2.9) for (any) given $\mathcal{G}_0^{-1}$. In order to do so, various kinds of technical methods can be applied that have been developed during the research on the Kondo problem. In particular, the physical intuition that has been obtained there can be translated into the properties of the Hubbard model. We now consider a fictitious Anderson model having the property that its effective action is given by (4.2.9)

$$H_{\mathrm{AM}} = \sum_{\boldsymbol{k},\sigma} \varepsilon_{\boldsymbol{k}} a^{\dagger}_{\boldsymbol{k}\sigma} a_{\boldsymbol{k}\sigma} - \frac{1}{\sqrt{N_0}}\left(V_{\boldsymbol{k}} a^{\dagger}_{\boldsymbol{k}\sigma} C_\sigma + V^*_{\boldsymbol{k}} C^{\dagger}_\sigma a_{\boldsymbol{k}\sigma}\right) + \varepsilon_d \sum_\sigma C^{\dagger}_\sigma C_\sigma + U n_\uparrow n_\downarrow\,. \tag{4.2.16}$$

Comparing with (3.1.22) and (4.1.1), notice that the electrons C and $C^\dagger$ correspond to $f, f^\dagger$ and $a, a^\dagger$ are fictitious conducting electrons. As demonstrated in Sect. 4.1, when a and $a^\dagger$ are integrated out, the effective action (4.2.9) is obtained, and $\mathcal{G}_0^{-1}(\mathrm{i}\omega_n)$ in this case is given by

$$\mathcal{G}_0^{-1}(\mathrm{i}\omega_n) = \mathrm{i}\omega_n - \varepsilon_d + \int_{-\infty}^{\infty} \frac{\mathrm{d}\varepsilon}{\pi} \frac{\Delta(\varepsilon)}{\mathrm{i}\omega_n - \varepsilon}, \tag{4.2.17}$$

$$\Delta(\varepsilon) = \frac{\pi}{N_0} \sum_{\boldsymbol{k}} |V_{\boldsymbol{k}}|^2 \delta(\varepsilon - \varepsilon_{\boldsymbol{k}}). \tag{4.2.18}$$

$\Delta(\varepsilon)$ is the function that represents the hybridization between the localized electrons and the conducting electrons. It is possible to express any kind of $\mathcal{G}_0^{-1}(\mathrm{i}\omega_n)$ using such a function $\Delta(\varepsilon)$. As follows from the discussion of Sect. 4.1, in the usual one-channel Kondo problem, under the assumption that $\Delta(0)$ is finite at the Fermi energy $\varepsilon = 0$, a Kondo peak with width of about T_{K} appears below a characteristic temperature (the Kondo temperature) T_{K} in the spectral function of the electron around $\varepsilon = 0$. Physically, this corresponds to singlet formation, when the mean value of the time scale of spin fluctuations exceeds the duration T_{K}^{-1}. The local Fermi liquid in this case means that the electrons are a Fermi liquid. On the other hand, when U is large, peaks at ε_{d} and $\varepsilon_{\mathrm{d}} + U$ arise, which correspond in this case to the lower and upper Hubbard bands, as will be explained in what follows.

Now, assuming that $\Delta(\varepsilon = 0)$ is finite, for $d \to \infty$ the model describes a Fermi liquid, and $\Delta(\varepsilon)$, or $\mathcal{G}_0^{-1}(\mathrm{i}\omega_n)$ have to be determined using the self-consistency condition. Because $\Delta(\varepsilon)$ is the electron spectrum at all other sites except the one that is considered, a possible self-consistent solution is given by a $\Delta(\varepsilon)$ having a gap, and likewise, a gap arises also in the electron spectrum. This corresponds to the Mott insulator solution discussed in Sect. 3.1, the solution for the half-filled case where U is larger than a critical value U_{c}. Therefore, the $d \to \infty$ model describes both the Fermi liquid phase case $U < U_{\mathrm{c}}$ and the Mott insulator phase $U > U_{\mathrm{c}}$, and provides a method to discuss the properties of the Mott transition. However, the validity of this model depends on whether the $\boldsymbol{k}$-dependence in the self-energy $\sum(\boldsymbol{k}, \omega)$ can be ignored; the reader should bear in mind that it has been claimed that some of the essential properties of the Mott transition are related just to this $\boldsymbol{k}$-dependence.

5. Gauge Theory of Strongly Correlated Electronic Systems

A powerful framework within which to study strongly correlated electronic systems is gauge theory. Different from the theories discussed so far, where the starting point has been the degrees of freedom on each site of the lattice, in this framework the theory is based on the degrees of freedom on the links connecting the sites. In this chapter, the gauge theory is presented for three examples, the quantum spin system, the doped quantum spin system, and the quantum Hall liquid.

5.1 Gauge Theory of Quantum Anti-ferromagnets

In Sect. 2.3, we saw that the Berry phase plays an important role in one-dimensional quantum anti-ferromagnets. We will now discuss this point in an even clearer formulation – using gauge theory. Because the gauge field is the phase factor arising in the inner product of quantum mechanical states – the so-called connection in mathematical language – it is intimately related to the Berry phase; one might even say it is the same thing. Let us work out this relation in terms of the non-linear sigma model that was introduced in Sect. 2.3.

In order to do so, we represent $\boldsymbol{\Omega}$ using the complex fields z_α ($\alpha =\uparrow, \downarrow$) as

$$\boldsymbol{\Omega}(x) = \sum_{\alpha,\beta} z^*_\alpha(x)\boldsymbol{\sigma}_{\alpha\beta} z_\beta(x) . \tag{5.1.1}$$

$\boldsymbol{\sigma}$ are the Pauli matrices $\boldsymbol{\sigma} = (\sigma^x, \sigma^y, \sigma^z)$ introduced in Sect. 1.1. Notice that this expression corresponds exactly to the expectation value of the spin direction of the two-component spin wave function:

$$\begin{bmatrix} z_\uparrow \\ z_\downarrow \end{bmatrix} . \tag{5.1.2}$$

However, in the present case, z_α should rather be understood as the integration variable in the path integral. Equation (5.1.1) equals

$$\boldsymbol{\Omega}(x)^2 = \left(|z_\uparrow(x)|^2 + |z_\downarrow(x)|^2\right)^2 \tag{5.1.3}$$

and the constraint $\boldsymbol{\Omega}(x)^2 = 1$ translates into

$$|z_\uparrow(x)|^2 + |z_\downarrow(x)|^2 = 1 \,. \tag{5.1.4}$$

Because $\boldsymbol{\Omega}$ has three components, and there is one constraint, the number of degrees of freedom is two. On the other hand, because $z_\uparrow$ and $z_\downarrow$ are complex fields with four degrees of freedom, there remain three degrees of freedom under the constraint (5.1.4). Therefore, the number of degrees of freedom differs by one. This remaining one corresponds to the gauge degree of freedom

$$\begin{aligned} z_\alpha(x) &\to \mathrm{e}^{\mathrm{i}\theta(x)} z_\alpha(x) \,, \\ z_\alpha^*(x) &\to \mathrm{e}^{-\mathrm{i}\theta(x)} z_\alpha^*(x) \,. \end{aligned} \tag{5.1.5}$$

Because $\boldsymbol{\Omega}$ in (5.1.1) is invariant under the transformation (5.1.5), the physics is invariant under this transformation. That is, the theory is gauge invariant under the gauge transformation (5.1.5).

We now express the action integral explicitly in terms of z_α. After some calculation, we are left with

$$\frac{1}{g}\int (\partial_\mu \boldsymbol{\Omega}(x))^2 \,\mathrm{d}^2x = \frac{1}{g}\int \left\{\partial_\mu z_\alpha^* \partial_\mu z_\alpha + (z_\alpha^* \partial_\mu z_\alpha)(z_\beta^* \partial_\mu z_\beta)\right\} \mathrm{d}^2x \,. \tag{5.1.6}$$

The second term of the integrand in (5.1.6) can be re-expressed using the Stratonovich–Hubbard transformation as

$$\begin{aligned} &\exp\left[-\frac{1}{g}\int \mathrm{d}^2x \,(z_\alpha^*\partial_\mu z_\alpha)(z_\beta^*\partial_\mu z_\beta)\right] \\ &= \int \mathcal{D}a_\mu \exp\left[-\frac{1}{g}\int \mathrm{d}^2x \,(a_\mu^2 - 2\mathrm{i}a_\mu z_\alpha^* \partial_\mu z_\alpha)\right] \\ &= \int \mathcal{D}a_\mu \exp\left[-\frac{1}{g}\int \mathrm{d}^2x \,(z_\alpha^* z_\alpha a_\mu^2 - 2\mathrm{i}a_\mu z_\alpha^* \partial_\mu z_\alpha)\right] . \end{aligned} \tag{5.1.7}$$

We used (5.1.4) in the second line.

We conclude that, using the covariant derivative $\partial_\mu + \mathrm{i}a_\mu$, we can write

$$\exp\left[-\frac{1}{g}\int \mathrm{d}^2x \,|\partial_\mu \boldsymbol{\Omega}(x)|^2\right] = \int \mathcal{D}a_\mu \exp\left[-\frac{1}{g}\int \mathrm{d}^2x \,|(\partial_\mu + \mathrm{i}a_\mu) z_\alpha|^2\right] . \tag{5.1.8}$$

a_μ is the gauge field, and completing the square in (5.1.7) leads to the relation

$$a_\mu = \mathrm{i} z_\alpha^* \partial_\mu z_\alpha \,. \tag{5.1.9}$$

We conclude that under the gauge transformation (5.1.5), the gauge field transforms as

$$a_\mu \to a_\mu - \partial_\mu \theta(x) \,. \tag{5.1.10}$$

It is obvious that the action (5.1.8) is gauge invariant under the gauge transformation given by (5.1.5) and (5.1.10).

The meaning of a_μ can be explained as follows. The inner product of the state $|\chi_1\rangle = [z_\uparrow(x), z_\downarrow(x)]^t$ at point x and the state $|\chi_2\rangle = [z_\uparrow(x+\Delta x), z_\downarrow(x+\Delta x)]^t$ is given by

$$\begin{aligned}\langle\chi_1|\chi_2\rangle &= z_\alpha^*(x) z_\alpha(x+\Delta x) \approx z_\alpha^*(x)[z_\alpha(x) + \Delta x_\mu \partial_\mu z_\alpha(x)] \\ &= 1 + \Delta x_\mu z_\alpha^* \partial_\mu z_\alpha \\ &= 1 - \mathrm{i}\Delta x_\mu a_\mu(x) \approx \mathrm{e}^{-\mathrm{i}\Delta x_\mu a_\mu(x)} .\end{aligned}$$

Therefore, a_μ is the connection between the two states. It is easy to imagine that there is an intimate relationsip to the Berry phase of the spin. Indeed, after some calculation, we obtain the relation

$$\tfrac{1}{2}\boldsymbol{\Omega}\cdot(\partial_\tau \boldsymbol{\Omega} \times \partial_x \boldsymbol{\Omega}) = \partial_\tau(z_\alpha^* \partial_x z_\alpha) - \partial_x(z_\alpha^* \partial_\tau z_\alpha) . \tag{5.1.11}$$

Therefore, the Berry term (2.3.22) can be expressed as

$$\mathrm{i}S \int \mathrm{d}^2 x \, (\partial_\tau a_x - \partial_x a_\tau) = S \int \mathrm{d}^2 x \, E_x . \tag{5.1.12}$$

Here, E_x is the 'electric field' in the x direction. Putting things together, finally, the action reads

$$A = \frac{1}{g} \int \mathrm{d}^2 x \, |(\partial_\mu + \mathrm{i} a_\mu) z_\alpha|^2 + S \int \mathrm{d}^2 x \, E_x . \tag{5.1.13}$$

This action is equivalent to the problem of a bosonic field where at both ends of the sample external charges of strength $\pm S$ are present. When S is an integer, the charges on both ends are screened by the particles represented by z_α, and we expect that inside the sample the electric field is zero. This corresponds to the Haldene state. On the other hand, when the charges S are half-odd-integer, the effect of the screening of the particles z can only shift the charge $-1/2$ to $1/2$, or $1/2$ to $-1/2$, therefore total screening is not possible. As can be imagined from the above discussion, even if the particles in the probe with charge ± 1 where to reverse the direction of the electric field between each partner of the pair by pair creation, this state would be energetically degenerate with respect to other configurations. This is a sharp contrast to the case $S = 1$, where in the ground state no electric field is present inside the probe, and the pairs are confined due to the energy increase proportional to the distance between the partners of each pair. Because the particle z has spin $1/2$ (as is obvious from (5.1.1) and (5.1.2)), this means that in the integer spin case, the spin $1/2$ particle does not appear, but is confined to a bound state with spin 0 or spin 1.

On the other hand, in the half-odd-integer spin case, owing to the degeneracy described above, the spin $1/2$ particle is not confined, but is an independent particle. It is the so-called spinon, corresponding to the kink that was discussed in Sect. 1.2 and 1.3. Furthermore, a striking prediction of the above theory is the presence of a localized spin 1/2 near each end of the sample for $S = 1$, which arises due to the screening gauge charge (z-particle) bound to the external charge. This has been confirmed theoretically and experimentally.

5.2 Gauge Theory of the Doped Mott Insulator

As was pointed out in Sect. 3.1, the anti-ferromagnetic quantum spin system turns out to be an effective model of the Mott insulator. In the previous section, we saw how the gauge field is related to the quantum spin system. We can imagine that for the case that carriers are doped into the system, an interaction arises between the gauge field and these carriers. Indeed, this model is intensively studied as a model for high-temperature superconductors, and in what follows we will present this theory, focusing on the principal ideas.

Again, we consider the Hubbard model. We introduce the Stratonovich-Hubbard transformation, and as was done in Sect. 3.3, we will not only consider the z-direction, but we will also perform a path integral with respect to the quantization direction. Doing so, the spin fluctuation field becomes a vector field $\boldsymbol{\varphi_i}$. Now, we consider the case where U is large compared with t. Then, the interaction of the electrons and the field $(U/2)\boldsymbol{\varphi_i}$ is extremely strong. Therefore, the spin of the electron

$$\boldsymbol{s}_i = \frac{1}{2}\sum_{\alpha,\beta} C^\dagger_{i\alpha}\boldsymbol{\sigma}_{\alpha\beta}C_{i\beta}$$

is forced to align antiparallel to the direction $\boldsymbol{\varphi}_i$. Once the spin $\boldsymbol{s}_i$ is aligned antiparallel to $\boldsymbol{\varphi}_i$, the effect of U is saturated, no matter if $U \to \infty$. The strong correlation limit corresponds exactly to this saturated case, where the spin component parallel to $\boldsymbol{\varphi}_i$ is energetically strongly unpreferred and can be ignored. (Here, we consider the case where holes are doped, that is, less than one electron is present at every site.) In this case, the spin degeneracy is effectively lifted, and in the half-filled case, one electron occupies one site, respectively, and the system is an insulator. This is nothing but the Mott insulator introduced in Sect. 3.1. For the case when the number of electrons per side is smaller than one, that is, when a ratio of x holes is doped, then electrons that are in the neighbourhood of a hole can transfer to this site. We now will discuss this process further.

We consider an electron sitting at site i with spin

$$-\boldsymbol{\varphi}_i // \boldsymbol{n}_i = (\cos\phi_i \sin\theta_i, \sin\phi_i \sin\theta_i, \cos\theta_i)\,.$$

The spin wave function $|\chi_i\rangle$ of this state is given by the two-component spinor

$$|\chi_i\rangle = \begin{bmatrix} \mathrm{e}^{\mathrm{i}(b_i+\phi_i)/2}\cos\frac{\theta_i}{2} \\ \mathrm{e}^{\mathrm{i}(b_i-\phi_i)/2}\sin\frac{\theta_i}{2} \end{bmatrix}\,. \tag{5.2.1}$$

Here, b_i is the total phase factor of the wave function, corresponding to the gauge degree of freedom. This factor has no influence on the physics, as was discussed in (5.1.5). Indeed, when the expectation value of the spin operator is constructed with this state, the result is

$$\langle \chi_i | \boldsymbol{s}_i | \chi_i \rangle = \tfrac{1}{2} \langle \chi_i | \boldsymbol{\sigma} | \chi_i \rangle = \tfrac{1}{2} \boldsymbol{n}_i \, . \tag{5.2.2}$$

On the other hand, when an electron transfers onto site i with $-\varphi_i \| n_i$, then the spin wave must become $|\chi_i\rangle$ – you must change your habits if you go abroad. As a result, the transfer matrix element from site j to i is the inner product of $|\chi_i\rangle$ and $|\chi_j\rangle$, multiplied by t appearing in the Hubbard model

$$t_{ij}^{\text{eff}} = t \langle \chi_i | \chi_j \rangle \, . \tag{5.2.3}$$

Explicitly, from (5.2.1) we obtain t_{ij}^{eff} as

$$t_{ij}^{\text{eff}} = t\, \mathrm{e}^{-\mathrm{i}(b_i - b_j)/2} \cdot \left(\mathrm{e}^{-\mathrm{i}(\phi_i - \phi_j)/2} \cos\frac{\theta_i}{2} \cos\frac{\theta_j}{2} + \mathrm{e}^{\mathrm{i}(\phi_i - \phi_j)/2} \sin\frac{\theta_i}{2} \sin\frac{\theta_j}{2} \right) . \tag{5.2.4}$$

The absolute value is given by

$$\begin{aligned} \left| t_{ij}^{\text{eff}} \right|^2 &= t^2 \left[\cos^2\frac{\theta_i}{2} \cos^2\frac{\theta_j}{2} + \sin^2\frac{\theta_i}{2} \sin^2\frac{\theta_j}{2} \right. \\ &\quad \left. + 2 \cos\frac{\theta_i}{2} \cos\frac{\theta_j}{2} \sin\frac{\theta_i}{2} \sin\frac{\theta_j}{2} \cos(\phi_i - \phi_j) \right] \\ &= \frac{t^2}{2} [1 + \cos\theta_{ij}] = t^2 \cos^2\frac{\theta_{ij}}{2} \, . \end{aligned} \tag{5.2.5}$$

Here, θ_{ij} is the angle between $\boldsymbol{n}_i$ and $\boldsymbol{n}_j$ ($\cos\theta_{ij} = \boldsymbol{n}_i \boldsymbol{n}_j$).

That is, $|t_{ij}^{\text{eff}}|$ is maximal and equals t when $\boldsymbol{n}_i$ and $\boldsymbol{n}_j$ are parallel, and is zero for the antiparallel case $\boldsymbol{n}_i = -\boldsymbol{n}_j$. Because the spin is conserved during the transfer of the electron, this is a natural result. Furthermore, because of the gain in kinetic energy, the interaction is ferromagnetic-like. This corresponds to the so-called doublet-exchange interaction. The above considerations are valid for the absolute value of t_{ij}^{eff}, and as can be seen from (5.2.4), t_{ij}^{eff} is in general complex, and therefore has a phase factor. Calling this phase $\mathrm{e}^{ia_{ij}}$, we obtain

$$t_{ij}^{\text{eff}} = t\, \mathrm{e}^{\mathrm{i}a_{ij}} \cos\frac{\theta_{ij}}{2} \, . \tag{5.2.6}$$

The meaning of a_{ij} becomes clear by fixing the gauge for b_i. We set $b_i = -\phi_i$. Writing $\theta_i = \theta_j + \mathrm{d}\theta$ and $\phi_i = \phi_j + \mathrm{d}\phi$, from (5.2.4) we obtain

$$t_{ij}^{\text{eff}} = t \left[1 + \mathrm{i}\, \mathrm{d}\phi \sin^2\frac{\theta_i}{2} \right] \tag{5.2.7}$$

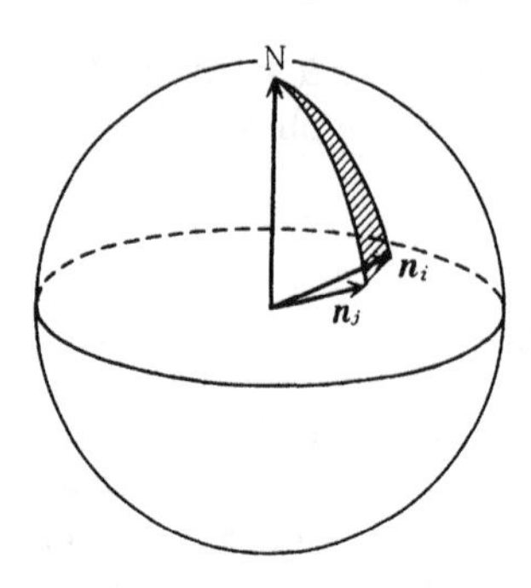

Fig. 5.1. Solid angle spanned by the north pole of the unit sphere and the two vectors $\boldsymbol{n}_i$ and $\boldsymbol{n}_j$

and therefore

$$a_{ij} = \mathrm{d}\phi \sin^2 \frac{\theta_i}{2} = \frac{\mathrm{d}\phi}{2}(1 - \cos\theta_i) .$$

As shown in Fig. 5.1, this is 1/2 times the surface of the triangle spanned by the north pole N, $\boldsymbol{n}_i$ and $\boldsymbol{n}_j$ on the unit sphere. This is also true in general. That is, a_{ij} is related to the solid angle spanned by the spin direction (the so-called spin chirality), and as was discussed in the previous section, the electrons feel its influence as 'magnetic field'.

Notice that in the limit $U \to \infty$, instead of the degree of freedom of the site, the degree of freedom t_{ij}^{eff} on the link becomes important. This point of view is qualitatively different from the discussion of the spin fluctuations in Sect. 3.4, or the dynamical mean field approach in Sect. 4.2. Especially spin singlets preserve the 'relation' between the spins, but because both spin states are fluctuating due to quantum effects, the degree of freedom on the link is the appropriate description. On the other hand, magnetic ordering can be described in an appropriate manner by the spin at each site. Both points of view are expected to be related to each other; however, at present, a unified description of both features has not been achieved. This problem remains to be solved in further studies.

In the limit of strong correlation, there exists another method for the derivation of the effective action than the one presented in the previous section. This method is called the slave-particle method (slave-boson method, slave-fermion method, see Sect. 4.1). In this framework, the creation and annihilation operators of the electron are represented in terms of the spinon $s_{i\sigma}^\dagger, s_{i\sigma}$, the holon $h_i^\dagger, h_i$, and the doublon $d_i^\dagger, d_i$:

$$\begin{aligned} C_{i\sigma}^\dagger &= s_{i\sigma}^\dagger h_i + \varepsilon_{\sigma\sigma'} s_{i\sigma'} d_i^\dagger , \\ C_{i\sigma} &= s_{i\sigma} h_i^\dagger + \varepsilon_{\sigma\sigma'} s_{i\sigma'}^\dagger d_i . \end{aligned} \tag{5.2.8}$$

Corresponding to the fact that the state at every site can be (a) empty, (b) $\uparrow$ spin, (c) $\downarrow$ spin or (d) double occupancy, the constraint reads

$$h_i^\dagger h_i + s_{i\uparrow}^\dagger s_{i\uparrow} + s_{i\downarrow}^\dagger s_{i\downarrow} + d_i^\dagger d_i = 1 . \tag{5.2.9}$$

Under this constraint, $C^\dagger_{i\sigma}$ and $C_{i\sigma}$ fulfil the usual fermion anti-commutation relations, when the spinon fulfils fermion (boson), the holon and the doublon fulfil boson (fermion) commutation/anti-commutation relations. This description is the so-called slave-boson (slave-fermion) method.

The coordinate transformation considered above would naturally lead to the slave-fermion method, however, in what follows, we will use the slave-boson picture. The reason is that this method is more appropriate to describe high-temperature superconductors. The advantage is that for the case when the repulsive force U between electrons at the same site is large, and double occupancy states therefore are forbidden, it is sufficient to ignore the doublon $d^\dagger_i, d_i$ in (5.2.8) and (5.2.9). Then, (5.2.8) becomes

$$\begin{aligned} C^\dagger_{i\sigma} &= s^\dagger_{i\sigma} h_i \,, \\ C_{i\sigma} &= s_{i\sigma} h^\dagger_i \,, \end{aligned} \tag{5.2.10}$$

and (5.2.9) becomes

$$h^\dagger_i h_i + s^\dagger_{i\uparrow} s_{i\uparrow} + s^\dagger_{i\downarrow} s_{i\downarrow} = 1 \,. \tag{5.2.11}$$

However, under this constraint, $C^\dagger_{i\sigma}$ and $C_{i\sigma}$ no longer fulfil fermion anti-commutation relations.

A model that describes the low-energy states, where the double occupancy states are forbidden, is the Hamiltonian of the t-J model that was introduced in Sect. 3.1

$$H = -\sum_{ij\sigma} t_{ij} C^\dagger_{i\sigma} C_{j\sigma} + \sum_{ij} J_{ij} \boldsymbol{S}_i \cdot \boldsymbol{S}_j \,. \tag{5.2.12}$$

In what follows, we asumme a high-temperature superconductor on a square lattice. By inserting (5.2.10), the Hamiltonian can be expressed in terms of $s^\dagger_{i\sigma}, s_{i\sigma}, h^\dagger_i$ and h_i. In the path integral, the Lagrangian for the t-J model becomes

$$\begin{aligned} L = &\sum_{i\sigma} s^\dagger_{i\sigma}(\partial_\tau - \mu_s) s_{i\sigma} + \sum_i h^\dagger_i (\partial_\tau - \mu_h) h_i \\ &- \sum_{ij,\sigma} t_{ij} s^\dagger_{i\sigma} s_{j\sigma} h_i h^\dagger_j + \frac{1}{3} \sum_{ij} J_{ij} s^\dagger_{i\alpha} \boldsymbol{\sigma}_{\alpha\beta} s_{i\beta} \cdot s^\dagger_{j\gamma} \boldsymbol{\sigma}_{\gamma\delta} s_{j\delta} \\ &+ \sum_i \lambda_i \left(h^\dagger_i h_i + \sum_\sigma s^\dagger_{i\sigma} s_{i\sigma} - 1 \right) . \end{aligned} \tag{5.2.13}$$

Here, the Lagrange multiplier field λ_i has been introduced in order to implement the constraint (5.2.11).

We now discuss the gauge invariance. Under the gauge transformation

$$s_{i\sigma} \to \mathrm{e}^{\mathrm{i}\varphi_i} s_{i\sigma} \,, \quad h_i \to \mathrm{e}^{\mathrm{i}\varphi_i} h_i$$

$C_{i\sigma}^\dagger$ and $C_{i\sigma}$ are invariant, and therefore the Hamiltonian is invariant. This gauge invariance gives rise to the appearance of the gauge field, which is also deeply connected to the constraint condition (5.2.9). In general, gauge fields often emerge in quantum field theories with constraints, and the present model is an explicit example.

We perform the Stratonovich–Hubbard transformation of the Lagrangian (5.2.13). For the calculation, we use the idendity

$$\begin{aligned} \boldsymbol{S}_i \cdot \boldsymbol{S}_j &= -\frac{1}{2}\Big(\sum_\sigma s_{i\sigma}^\dagger s_{j\sigma}\Big)\Big(\sum_\sigma s_{j\sigma}^\dagger s_{i\sigma}\Big) - \frac{1}{4} \\ &= -\frac{1}{2}\left(s_{i\uparrow}s_{j\downarrow} - s_{i\downarrow}s_{j\uparrow}\right)\left(s_{j\downarrow}^\dagger s_{i\uparrow}^\dagger - s_{j\uparrow}^\dagger s_{i\downarrow}^\dagger\right) + \frac{1}{4} \,. \end{aligned} \tag{5.2.14}$$

We insert this term into the expression proportional to J_{ij} in (5.2.12). In this case, there are two different ways to express the interaction, and presuming the mean field theoretical approach, both expressions must be summed as total action without factor 1/2. Doing so, and introducing the Stratonovich–Hubbard fields $\chi_{ij}, \bar{\chi}_{ij}, B_{ij}, \bar{B}_{ij}, \Delta_{ij}$ and $\bar{\Delta}_{ij}$, the action becomes

$$\begin{aligned} A = \int_0^\beta \mathrm{d}\tau \sum_{i,j} &\frac{1}{t_{ij}}\left(\bar{B}_{ij}\chi_{ij} + \bar{\chi}_{ij}B_{ij}\right) - \frac{J_{ij}}{2t_{ij}^2}\bar{\chi}_{ij}\chi_{ij} + \frac{1}{2J_{ij}}\bar{\Delta}_{ij}\Delta_{ij} \\ &+ \bar{\chi}_{ij}\sum_\sigma s_{i\sigma}^\dagger s_{j\sigma} + \chi_{ij}\sum_\sigma s_{j\sigma}^\dagger s_{i\sigma} \\ &+ \bar{\Delta}_{ij}\left(s_{i\uparrow}s_{j\downarrow} - s_{i\downarrow}s_{j\uparrow}\right) + \Delta_{ij}\left(s_{j\downarrow}^\dagger s_{i\uparrow}^\dagger - s_{j\uparrow}^\dagger s_{i\downarrow}^\dagger\right) \\ &+ \bar{\chi}_{ij}h_i^\dagger h_j + \chi_{ij}h_j^\dagger h_i \\ &+ \sum_i\Big\{\sum_\sigma s_{i\sigma}^\dagger(\partial_\tau + \lambda_i)s_{i\sigma} + h_i^\dagger(\partial_\tau + \lambda_i - \tilde{\mu}_{\mathrm{B}})h_i\Big\} \,. \end{aligned} \tag{5.2.15}$$

Here, $s^\dagger$ and $h^\dagger$ are the Grassmann and c-number corresponding to the original operators $s^\dagger$ and $h^\dagger$, respectively. χ_{ij} and Δ_{ij} correspond to the hopping and the singlet formation of the spinon (fermion), respectively. Correspondingly, the hopping of the holon (boson) is represented by B_{ij}. At the same time, the possibility that the boson itself condenses ($\langle h_i\rangle \neq 0$) has to be taken into account. Mean field approximation corresponds to determining the saddle-point solution of the action (5.2.15). Figure 5.2 shows the resulting schematic phase diagram in the temperature–hole density plane.

The highest temperature phase is the state dominated by incoherent hopping caused by thermal diffusion. Here, all order parameters are zero. Decreasing the temperature, the phase of uniform RVB states emerges, characterized by quantum mechanical motion of the spinons and the holons, and the spinons are Fermi degenerated. Here, $\chi_{ij}, \bar{\chi}_{ij}, B_{ij}$ and $\bar{B}_{ij}$ reach a constant value different from zero, and $\Delta_{ij} = \langle h_i\rangle = 0$.

Lowering the temperature, the uniform RVB state becomes unstable with respect to two kinds of order. One is the singlet pairing Δ_{ij} due to the J_{ij}

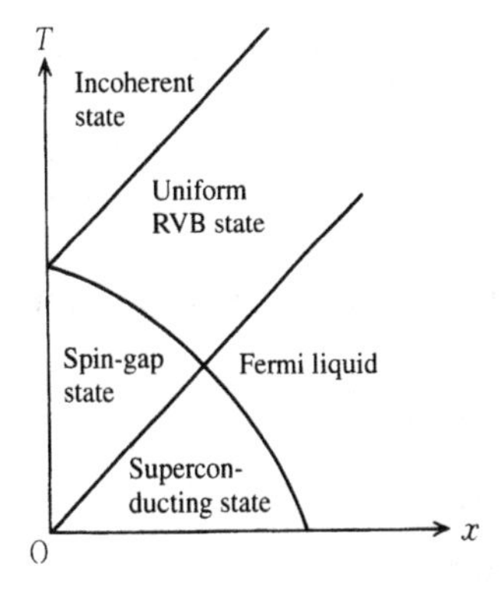

Fig. 5.2. The mean field approximation with the slave-boson method

of the spinons, the other is the Bose condensation of the holons. When only the first one occurs, the so-called spin-gap state emerges. On the other hand, if only the second, the Bose condensation, occurs, then the state becomes a normal Fermi liquid. The system becomes superconducting when both features occur at the same time, as can also be understood in principle from $C_{i\uparrow}C_{j\downarrow} = f_{i\uparrow}f_{j\downarrow}b_i^\dagger b_j^\dagger$.

The properties of the different states of the phase diagram presented above are under investigation; here, we will present an analysis of the uniform RVB phase in the framework of gauge theory. In this state, the spinons are moving in a band as in the normal tight-binding model and create a large Fermi surface; however, the holons are distributed according to the Boltzmann statistics, having only small momentum. The effective Lagrangian for energies smaller than J in the continuum limit becomes

$$
\begin{aligned}
L = & \int \mathrm{d}r \Bigg[s_\sigma^*(r,\tau)\left(\frac{\partial}{\partial\tau} - a_0 - \mu_\mathrm{F}\right) s_\sigma(r,\tau) \\
& + s_\sigma^*(r,\tau)\frac{1}{2m_\mathrm{F}}\left(\frac{1}{\mathrm{i}}\nabla + \boldsymbol{a}\right)^2 s_\sigma(r,\tau)\Bigg] \\
& + \int \mathrm{d}r \Bigg[h^*(r,\tau)\left(\frac{\partial}{\partial\tau} - a_0 - \mu_\mathrm{B}\right) h(r,\tau) \\
& + h^*(r,\tau)\frac{1}{2m_\mathrm{B}}\left(\frac{1}{\mathrm{i}}\nabla + \boldsymbol{a}\right)^2 h(r,\tau)\Bigg] .
\end{aligned}
\tag{5.2.16}
$$

Here, we used the units $\hbar = c = 1$, and the lattice constant is set to 1. Then, the exchange interaction J is measured in units of energy, m_B and m_F are of order 1, the Fermi temperature T_F of the spinon is of order $(1 - x) \simeq 1$, and the Bose condensation temperature $T_\mathrm{B.E.}^{(0)}$ of the holon is of order x (In the ideal two-dimensional Bose Gas, bose condensation at a finite temperature does not emerge. In this case, $T_\mathrm{B.E.}^{(0)}$ corresponds to the crossover temperature, where the absolute value of the chemical potential changes from a behaviour $\propto T$ to an $\exp[-T_\mathrm{B.E.}^{(0)}/T]$ behaviour).

Now, the gauge field $\boldsymbol{a}$ occurring in (5.2.16) corresponds to the fluctuations of the phase of the order parameters χ_{ij} and B_{ij}. Therefore, two phases should exist, and from (5.2.15) we conclude that only the in-phase mode is massless. At low energies, we assume that the out-of-phase mode can be ignored. This in-phase mode is the space component of the gauge field a_{ij} defined on the link between site i and site j, and in the continuum limit, it corresponds to the vector field $\boldsymbol{a}(r,\tau)$ expressed as

$$a_{ij} = (\boldsymbol{r}_i - \boldsymbol{r}_j) \cdot \boldsymbol{a}\left(\frac{\boldsymbol{r}_i + \boldsymbol{r}_j}{2}, \tau\right) . \tag{5.2.17}$$

On the other hand, the time component a_0 is the fluctuation of the Lagrange multiplier $\lambda_i(\tau)$ around the expectation value λ_0, and in the continuum limit, it is given by $\mathrm{i}a_0(r,\tau)$.

Next, we consider the coupling of the system to the electromagnetic field (expressed in terms of the vector potential A_0, A_{ij}). A_{ij} appears on scene as a phase factor in the transfer of the electron

$$t_{ij} C_{i\sigma}^\dagger C_{j\sigma} \to t_{ij}\,\mathrm{e}^{\mathrm{i}A_{ij}} C_{i\sigma}^\dagger C_{j\sigma} = t_{ij}\,\mathrm{e}^{\mathrm{i}A_{ij}} s_{i\sigma}^\dagger s_{j\sigma} h_i h_j^\dagger . \tag{5.2.18}$$

Because this phase factor can be interpreted as being multiplied by $s_{i\sigma}^\dagger s_{j\sigma}$, it is sufficient to replace the first term in (5.2.16) by

$$\frac{1}{2m_{\mathrm{F}}}\left(\frac{1}{\mathrm{i}}\nabla + \boldsymbol{a}\right)^2 \to \frac{1}{2m_{\mathrm{F}}}\left(\frac{1}{\mathrm{i}}\nabla + \boldsymbol{a} + \boldsymbol{A}\right)^2 . \tag{5.2.19}$$

The coupling of the scalar potential A_0 to the charge can be expressed as

$$A_0(i) \cdot \sum_\sigma C_{i\sigma}^\dagger C_{i\sigma} = A_0(i) \cdot \sum_\sigma f_{i\sigma}^\dagger f_{i\sigma} . \tag{5.2.20}$$

Notice that the above considerations would not be altered if the holon were charged. However, as will become clear in a moment, owing to the composition rule, both pictures are connected by a shift of the origin of the gauge field a_μ, and after the gauge field is integrated out, the physics is identical. Therefore, it makes no difference whether we assume s or h to be charged, because finally only gauge invariant expressions are physically meaningful.

The Zeeman coupling of the spin and the magnetic field can be expressed using only s and $s^\dagger$. This is due to the fact that in

$$\begin{aligned} \boldsymbol{s}_i &= \tfrac{1}{2} C_{i\alpha}^\dagger \boldsymbol{\sigma}_{\alpha\beta} C_{i\beta} = \tfrac{1}{2} f_{i\sigma}^\dagger \boldsymbol{\sigma}_{\alpha\beta} f_{i\beta} b_i b_i^\dagger \\ &= \tfrac{1}{2} f_{i\alpha}^\dagger \boldsymbol{\sigma}_{\alpha\beta} f_{i\beta}\left[1 - b_i^\dagger b_i\right] \end{aligned}$$

when $f_{i\beta}$ acts on the state $b_i^\dagger b_i = 1$, the result vanishes, and therefore in the above equation, $b_i^\dagger b_i$ can be ignored. That is, the degree of freedom of the spin only belongs to the fermions.

Writing the action (5.2.16) including the coupling to the electromagnetic field in a general manner, the result is

$$A_{\text{eff}} = A_{\text{s}}(\partial_\mu + \mathrm{i}a_\mu + \mathrm{i}A_\mu) + A_{\text{h}}(\partial_\mu + \mathrm{i}a_\mu)\,. \tag{5.2.21}$$

Differentiating with respect to a_μ, we obtain

$$\frac{\delta A_{\text{eff}}}{\delta a_\mu} = j_{\text{s}\mu} + j_{\text{h}\mu} = 0\,. \tag{5.2.22}$$

This means that the number of spinons plus the number of holons at every point is conserved. This is nothing but the constraint (5.2.11). That is, as shown in Fig. 5.3, the cancellation of the spinon current and the holon current simply means that the total number of spinon and holon is conserved at each site.

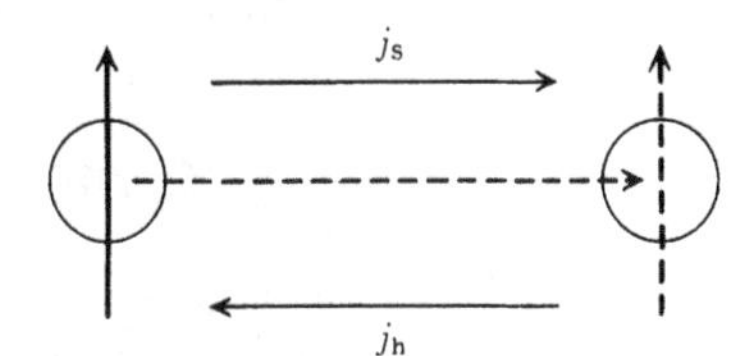

Fig. 5.3. Back flow effect: the flow of a spinon from the left to the right just signifies the flow of a holon (*hole*) from the right to the left, therefore $j_{\text{s}} + j_{\text{h}} = 0$ holds

The advantage of the gauge theory is the possibility to first integrate out the spinons and the holons and to determine the effective action of the gauge field. Explicitly, with the current–current correlation function $\Pi^{\text{s}}_{\mu\nu}(\boldsymbol{q},\omega_n)$ $(\Pi^{\text{h}}_{\mu\nu}(\boldsymbol{q},\omega_n))$ of the spinons (holons) up to second order in a_μ and A_μ we obtain

$$\begin{aligned} A_{\text{eff}}(a_\mu, A_\mu) &= \sum_{\boldsymbol{q},\omega_n} \Pi^{\text{h}}_{\mu\nu}(\boldsymbol{q},\omega_n) a_\mu(\boldsymbol{q},\omega_n) a_\nu(-\boldsymbol{q},-\omega_n) \\ &+ \sum_{\boldsymbol{q},\omega_n} \Pi^{\text{s}}_{\mu\nu}(\boldsymbol{q},\omega_n)(a_\mu(\boldsymbol{q},\omega_n) + A_\mu(\boldsymbol{q},\omega_n))\left(a_\nu(-\boldsymbol{q},-\omega_n) + A_\nu(-\boldsymbol{q},-\omega_n)\right). \end{aligned} \tag{5.2.23}$$

In the Coulomb gauge $(\nabla\cdot\boldsymbol{a} = 0, \nabla\cdot\boldsymbol{A} = 0)$, (5.2.23) can be expressed in terms of the scalar component Π_0 and the transversal component $\Pi_\perp$ as (using short notation)

$$\begin{aligned} A_{\text{eff}}(a_\mu, A_\mu) = &\sum_{\boldsymbol{q},\omega_n} \left\{\Pi^{\text{h}}_0 a_0 a_0 + \Pi^{\text{s}}_0 (a_0 + A_0)(a_0 + A_0)\right\} \\ &+ \sum_{\boldsymbol{q},\omega_n} \left\{\Pi^{\text{h}}_\perp a_\perp a_\perp + \Pi^{\text{s}}_\perp (a_\perp + A_\perp)(a_\perp + A_\perp)\right\}. \end{aligned} \tag{5.2.24}$$

From this expression, the following conclusions can be drawn.

1. The propagator describing the fluctuations in the gauge field is given by $D_\alpha(\boldsymbol{q},\omega_n) = (\Pi^{\text{s}}_\alpha + \Pi^{\text{h}}_\alpha)^{-1}$ $(\alpha = 0, \perp)$.

2. When a_μ is integrated out, the obtained effective action in terms of A_μ describes the electrical response of the system

$$A_{\text{eff}}(A_\mu) = \sum_{\alpha=0,\perp} \sum_{\boldsymbol{q},\omega_n} \Pi_\alpha(\boldsymbol{q},\omega_n) A_\alpha(\boldsymbol{q},\omega_n) A_\alpha(-\boldsymbol{q},-\omega_n), \quad (5.2.25)$$

$$\Pi_\alpha(\boldsymbol{q},\omega_n) = \frac{\Pi^{\text{s}}_\alpha(\boldsymbol{q},\omega_n)\Pi^{\text{h}}_\alpha(\boldsymbol{q},\omega_n)}{\Pi^{\text{s}}_\alpha(\boldsymbol{q},\omega_n) + \Pi^{\text{h}}_\alpha(\boldsymbol{q},\omega_n)}. \quad (5.2.26)$$

Therefore, (5.2.26) is the current–current correlation of the total system.

The meaning of these equations is the following. First, owing to the external field A_μ, the mean value of the fluctuations of a_μ is shifted about $-\Pi^{\text{s}}_\alpha A_\alpha/(\Pi^{\text{s}}_\alpha + \Pi^{\text{h}}_\alpha)$ and the mean value of the gauge fields that spinons and holons feel are given by $\Pi^{\text{h}}_\alpha A_\alpha/(\Pi^{\text{s}}_\alpha + \Pi^{\text{h}}_\alpha)$ and $-\Pi^{\text{s}}_\alpha A_\alpha/(\Pi^{\text{s}}_\alpha + \Pi^{\text{h}}_\alpha)$, respectively. Because of this screening effect, equation (5.2.22) can be fulfilled, and finally the composition law of the electrical resistance in a sequential circuit

$$\Pi_\alpha^{-1} = (\Pi^{\text{s}}_\alpha)^{-1} + (\Pi^{\text{h}}_\alpha)^{-1} \quad (5.2.27)$$

is fulfilled. This equation contains the information that the spinon and the holon cannot move independently in one vacant site, but depend on each other; that is, the slower of both partners Π^{s} and Π^{h} is dominating the velocity, and determines Π.

Because in the uniform RVB state the low-frequency, long-wavelength fluctuations of $a_\perp$ are important, we consider $\Pi^{\text{s}}_\perp$ and $\Pi^{\text{h}}_\perp$ in this case. With v being the characteristic velocity of the spinon (holon), in the region $|\omega_n| \ll v\,|\boldsymbol{q}|$, we write the expression

$$\Pi^{\text{s(h)}}_\perp(\boldsymbol{q},\omega_n) = |\omega_n|\sigma_{\text{s(h)}}(\boldsymbol{q}) + \chi_{\text{s(h)}}|\boldsymbol{q}|^2. \quad (5.2.28)$$

Here, $\chi_{\text{s(h)}}$ is the Landau diamagnetic susceptibility of the spinon (holon), and $\sigma_{\text{s(h)}}(\boldsymbol{q})$ is the wave number depending on the static conductivity. When x is smaller than one, $\sigma_{\text{s}}(q)$ is larger than $\sigma_{\text{h}}(q)$, and it is sufficient to consider only this conductivity. With l being the mean free path of the spinon, we obtain

$$\sigma_{\text{s}}(q) \approx \begin{cases} k_{\text{F}}/q & (ql > 1) \\ k_{\text{F}}l & (ql < 1) \end{cases},$$

and finally, the propagator of the gauge field $D_\perp(\boldsymbol{q},\omega_n)$ becomes

$$D_\perp(\boldsymbol{q},\omega_n) = \frac{1}{|\omega_n|\sigma_{\text{s}}(q) + \chi_{\text{d}}q^2}.$$

Here, the relation $\chi_{\text{d}} = \chi_{\text{s}} + \chi_{\text{h}}$ holds.

Next, we discuss the behaviour of the spinons and holons interacting with these gauge field fluctuations. We calculate the inelastic scattering rate τ^{-1} due to the scattering by the gauge field. In the framework of the Boltzmann theory, the transport scattering time $\tau_{\boldsymbol{k}}$ of a holon (boson) with wave number $\boldsymbol{k}$ is given by

$$\tau_{\boldsymbol{k}}^{-1}=\int \mathrm{d}^2\boldsymbol{q}\mathrm{d}\omega\left(\frac{q}{k}\right)^2\left(\frac{\boldsymbol{q}}{q}\times\frac{\boldsymbol{k}}{m_{\mathrm{B}}}\right)^2\frac{1}{\mathrm{e}^{\beta\omega}-1}\frac{\omega\sigma_{\mathrm{s}}(q)}{(\omega\sigma_{\mathrm{s}}(q))^2+(\chi_{\mathrm{d}}q^2)^2}\delta(\varepsilon_{\boldsymbol{k}}-\varepsilon_{\boldsymbol{k}+\boldsymbol{q}}+\omega)\,. \tag{5.2.29}$$

Here, $\varepsilon_{\boldsymbol{k}}=k^2/2m_{\mathrm{B}}-\mu_{\mathrm{B}}$ is the energy of the boson. The term $(q/k)^2$ in the integrand corresponds to the term $1-\cos\theta$ appearing in the calculation of the transport scattering rate. It signifies that when the change q of the wave number due to the scattering is small, the contribution to the relaxation of the boson wave number (the momentum) is small.

We will determine (5.2.29) explicitly. It can be self-consistently shown that the dominant contribution to the q-integral is given by the region $ql>1$. We can set $\sigma_{\mathrm{s}}(q)\simeq k_{\mathrm{F}}/q$. From the form of the spectral function of the gauge field, the integral is dominated by the region $|\omega|\simeq\chi_{\mathrm{d}}q^3/k_{\mathrm{F}}$. On the other hand, the wave number k of the boson is of magnitude $\sqrt{T}$, and from the energy conservation law $\delta(\varepsilon_{\boldsymbol{k}}-\varepsilon_{\boldsymbol{k}+\boldsymbol{q}}+\omega)$ we deduce that q also is of magnitude $\sqrt{T}$. We conclude that $|\omega|\simeq T^{3/2}\ll T$ (we consider the case $T_{\mathrm{B.E.}}<T\ll J$), and we ignore ω in the δ function compared with $\varepsilon_{\boldsymbol{k}}-\varepsilon_{\boldsymbol{k}+\boldsymbol{q}}\simeq T$. The Bose factor $(\mathrm{e}^{\beta\omega}-1)^{-1}$ is approximated by $(\beta\omega)^{-1}=T/\omega$. Then, the ω integration can be performed, and we are left with

$$\tau_{\boldsymbol{k}}^{-1}\approx\int \mathrm{d}^2\boldsymbol{q}\,\frac{1}{k^2}\left(\frac{\boldsymbol{q}\times\boldsymbol{k}}{m_{\mathrm{B}}}\right)^2\frac{T}{\chi_{\mathrm{d}}}\frac{1}{q^2}\delta\left(\frac{2\boldsymbol{k}\cdot\boldsymbol{q}+q^2}{2m_{\mathrm{B}}}\right)\sim\frac{T}{\chi_{\mathrm{d}}m_{\mathrm{B}}}=\tau_{\mathrm{B}}^{-1}\,. \tag{5.2.30}$$

In such a way, the characteristic temperature dependence of the low-frequency, long-wavelength fluctuations can be deduced. A similar calculation for the spinons (fermions) leads to the result $\tau_{\mathrm{F}}^{-1}\simeq J(T/J)^{4/3}$. Owing to Fermi degeneracy, the fermions have a large kinetic energy of magnitude of the Fermi energy, and the coupling to the gauge field becomes weaker. Therefore, the relaxation ratio becomes smaller.

With these results, using the composition law mentioned in (5.2.27), we can determine the conductivity of the whole system. Because the conductivity $\sigma(\omega)$ is given by $\sigma(\omega)\propto\Pi_{\perp}(\omega,\boldsymbol{q}=0)/\omega$, from (5.2.27) we obtain

$$\sigma^{-1}=\sigma_{\mathrm{s}}^{-1}+\sigma_{\mathrm{h}}^{-1}\,.$$

From the above considerations, since the resistivity of the holons ρ_{h} is larger than ρ_{s}, we obtain

$$\rho=\rho_{\mathrm{s}}+\rho_{\mathrm{h}}\cong\rho_{\mathrm{h}}\sim T/x\,. \tag{5.2.31}$$

This could be one scenario explaining the experiments. This composition law can be generalized also for other physical quantities besides the resistivity. The analysis of the Hall coefficient, thermal conductivity, thermopower, magnetoresistance, photoemission spectra, NMR and neutron scattering experiments have been performed. It is important to notice that the role of the spinons and holons appear in the different physical quantities in a different light. Besides the thermal conductivity, in all cases the transport coefficient is dominated by the holons; and shows the doped semiconductor-like characteristics, on the other hand, the magnetic properties, as visible in NMR and neutron scattering, are dominated by the spinons. Corresponding to (5.2.10), the photoemission spectrum of photo electrons is given by the convolution integral of the Green functions of the spinons and the holons, and a large Fermi sea of spinons can be observed. This kind of double-faced description is in good correspondence with experiments.

5.3 Gauge Theory of Quantum Hall Liquids

Non-local interactions can be expressed in terms of gauge fields with a local interaction. Especially in the two-dimensional electron system, the Chern–Simons gauge field plays a crucial role in quantum Hall liquids because it transmutes the statistics of the particles. The principal idea can be seen in Fig. 5.4.

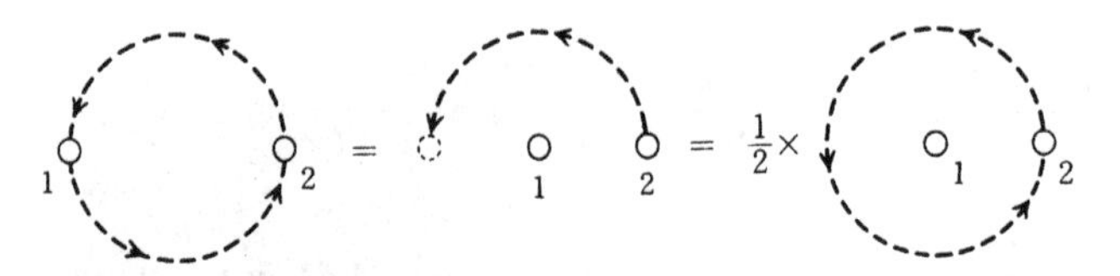

Fig. 5.4. The exchange of two particles and the Aharonov–Bohm effect

The statistical factor arising when two particles are exchanged is given by the phase factor $e^{i\theta}$. $\theta = 2\pi n$ corresponds to bosons, and $\theta = \pi(2n+1)$ to fermions. When the system is time- or space-inversion invariant, $e^{i\theta} = e^{-i\theta}$ can be deduced, and the statistics are restricted to either the fermionic or the bosonic case. However, for the quantum Hall system, where an external magnetic field is present, particles with θ different from $2\pi n$ and $2\pi(n+1)$ may exist. These particles are called anyons.

A similar phase factor also arises in the so-called Aharonov–Bohm effect. That is, when one charged particle turns once around one flux string, the phase factor

$$\frac{e}{\hbar c}\oint_C \boldsymbol{A}\cdot \mathrm{d}\boldsymbol{r} = 2\pi\frac{\Phi}{\phi_0} \tag{5.3.1}$$

arises. $\phi_0 = hc/e$ is the gauge flux quantum (unit flux). Φ is the magnetic flux going through the surface with boundary C. As demonstrated in Fig. 5.4, the interchange of two particles is equivalent to half the rotation of particle 2 around particle 1. By associating the flux θ/π for both particles, the phase factor $\mathrm{e}^{\mathrm{i}\theta}$ arises when the two particles are exchanged. When the original statistical phase factor of the particle is $\mathrm{e}^{\mathrm{i}\theta_0}$, then this factor alters to $\mathrm{e}^{\mathrm{i}(\theta_0+\theta)}$, and in this way it is possible to change the statistics of the particle.

Explicitly, the 'associated flux' can be realized as the Chern–Simons gauge field a_μ. That is, $\boldsymbol{a}$ is determined by the conditions

$$\nabla \times \boldsymbol{a} = 2\theta\rho(\boldsymbol{r}) \tag{5.3.2}$$

and

$$\nabla \cdot \boldsymbol{a} = 0\,, \tag{5.3.3}$$

with $\rho(\boldsymbol{r})$ being the particle density.

It is sufficient to replace the derivative $\mathrm{i}\nabla$ in the particle Hamiltonian by the covariant derivative $\mathrm{i}\nabla + \boldsymbol{a}$. The condition (5.3.2) can by implemented by using the Lagrange multiplier field a_0 and adding the term

$$\int_0^\beta \mathrm{d}\tau \int \mathrm{d}\boldsymbol{r}\, a_0 \left(\frac{1}{2\theta} \nabla \times \boldsymbol{a} - \rho \right) \tag{5.3.4}$$

to the action. In a gauge invariant manner, the first term in the integrand of (5.3.4) must be expressed as

$$A_{\mathrm{C.S.}} = \int_0^\beta \mathrm{d}\tau \int \mathrm{d}\boldsymbol{r} \frac{1}{4\theta} \varepsilon^{\mu\nu\lambda} a_\mu \partial_\nu a_\lambda\,. \tag{5.3.5}$$

$\varepsilon^{\mu\nu\lambda}$ is the totally antisymmetric tensor, with $\varepsilon^{012} = 1$. Writing the action of the particle as $A_{\mathrm{particle}}(\psi^\dagger, \psi, \partial_\mu)$, then the total action corresponding to (5.2.21) becomes

$$A = A_{\mathrm{C.S.}}(a_\mu) + A_{\mathrm{particle}}(\psi^\dagger, \psi, \partial_\mu + \mathrm{i}a_\mu + \mathrm{i}A_\mu)\,. \tag{5.3.6}$$

Here, $\psi^\dagger$ and ψ are 'composite fields' that are the combined state of one electron with one gauge flux.

The statistics of this composite particle are determined by θ, which cannot be uniquely defined. First, we consider the case where the mean value of the gauge flux $\nabla \times \boldsymbol{a}$, which is given in terms of the mean value $\bar{\rho}$ of the electron density by

$$\langle \nabla \times \boldsymbol{a} \rangle = 2\theta\bar{\rho}\,,$$

just cancels with the external field $\boldsymbol{B} = \nabla \times \boldsymbol{A}$. This condition is given by

$$2\theta\bar{\rho} + B = 0\,. \tag{5.3.7}$$

Recall that the filling factor ν of the Landau level is given by

$$\nu = \frac{\bar{\rho}}{G} = \frac{\bar{\rho}}{|B|/2\pi}, \tag{5.3.8}$$

where G is the degeneracy per unit area of the Landau level. We conclude that it is sufficient to choose

$$\theta = \pi\nu^{-1} \tag{5.3.9}$$

in order to fulfil the condition (5.3.7). In the case $\nu = 1/(2m+1)$, θ is an odd multiple of π, and therefore the composite field is a boson. In the mean field approximation, bosonic composite particles condense when no external magnetic field is present and become superfluid. This scenario simply describes the coherence in the quantum Hall liquid, as was described in [G1]. Here, we discuss a different point of view, related to composite fermions, in more detail.

Returning to (5.3.9) and considering the case $\nu = 1/2$, the obtained composite particle again becomes a fermion, because θ is then given by $\theta = 2\pi$. Because this composite particle on average does not feel the influence of the magnetic field, we expect that the state can be described as a Fermi liquid. Furthermore, the mean value of the magnetic field ΔB that the composite fermion with flux $\theta = 2\pi$ feels for general ν is given by

$$\Delta B = |B| - 4\pi\bar{\rho} = 2\pi\bar{\rho}\left(\frac{1}{\nu} - 2\right), \tag{5.3.10}$$

Therefore, the filling factor $\nu_{\mathrm{C.F.}}$ of the Landau level of the composite fermions that feel ΔB is given by

$$\nu_{\mathrm{C.F.}}^{-1} = \frac{\Delta B}{2\pi\bar{\rho}} = \nu^{-1} - 2. \tag{5.3.11}$$

The integer quantum Hall effect of the composite fermions only occurs when the Landau level is totally filled, that is, when $\nu_{\mathrm{C.F.}}$ is an integer number p. Rewriting this in terms of ν, we obtain

$$\nu = \nu_p = \frac{1}{\dfrac{1}{p} + 2} = \frac{p}{2p+1}. \tag{5.3.12}$$

The series of these filling factors is in good agreement with the observed plateaux in experiments. Furthermore, $\nu = 1/2$ can be considered as the limiting value for $p \to \infty$.

Furthermore, when composite fermions with general $\theta = 2\pi n$ (n; integer) are constructed, instead of (5.3.12) we obtain the series

$$\nu_{n,p} = \frac{p}{2np+1}. \tag{5.3.13}$$

For such a series, there exists a gap in the excitation spectrum of the composite fermions. This gap E_{G} is proportional to ΔB and anti-proportional to the effective mass m^* of the composite fermions m^*:

$$E_{\mathrm{G}} \sim \frac{\Delta B}{m^*} = \frac{2\pi\bar{\rho}}{m^*}(\nu_{n,p}^{-1} - 2n) = \frac{2\pi\bar{\rho}}{m^* p} . \tag{5.3.14}$$

We should stress that m^* is a somewhat peculiar object. Naively, the reader might think that m^* is given by the band mass m_{b} of the electron without a magnetic field. However, due to the magnetic field, Landau levels emerge, leading to a state where the kinetic energy of the electrons is quenched. Especially when the lowest-lying Landau level is considered, the only energy scale is given by the Coulomb energy

$$E_{\mathrm{C}} = \frac{e^2}{\varepsilon l_B} . \tag{5.3.15}$$

Here, ε is the dielectric constant, and l_B is the radius of the cyclotron motion $l_B \sim B^{-1/2}$. Because this energy corresponds to the kinetic energy with effective mass m^*, we obtain the estimation

$$\frac{\hbar^2}{m^* l_B^2} \sim \frac{e^2}{\varepsilon l_B} . \tag{5.3.16}$$

Next, we consider the fluctuations around the mean value. With the same technique as in the previous section, we expand the action (5.3.6) up to second order in $\delta a_\mu + \delta A_\mu = a_\mu + A_\mu - \langle a_\mu \rangle - \bar{A}_\mu$;

$$\begin{aligned} A_{\mathrm{eff}} &= \sum_{\boldsymbol{q},\mathrm{i}\omega_n} \begin{bmatrix} \delta a_x \\ \delta a_y \end{bmatrix}_{(-\boldsymbol{q},-\mathrm{i}\omega_n)} \begin{bmatrix} 0 & -\dfrac{\mathrm{i}\omega_n}{4\theta} \\ \dfrac{\mathrm{i}\omega_n}{4\theta} & 0 \end{bmatrix} \begin{bmatrix} \delta a_x \\ \delta a_y \end{bmatrix}_{(\boldsymbol{q},\mathrm{i}\omega_n)} \\ &+ \frac{1}{2}\sum_{\boldsymbol{q},\mathrm{i}\omega_n} \begin{bmatrix} \delta a_x + \delta A_x \\ \delta a_y + \delta A_y \end{bmatrix}_{(-\boldsymbol{q},-\mathrm{i}\omega_n)} \begin{bmatrix} \Pi_{xx}^{\mathrm{CF}} & \Pi_{xy}^{\mathrm{CF}} \\ \Pi_{yx}^{\mathrm{CF}} & \Pi_{yy}^{\mathrm{CF}} \end{bmatrix}_{(\boldsymbol{q},\mathrm{i}\omega_n)} \begin{bmatrix} \delta a_x + \delta A_x \\ \delta a_y + \delta A_y \end{bmatrix}_{(\boldsymbol{q},\mathrm{i}\omega_n)} . \end{aligned} \tag{5.3.17}$$

Here, we chose the gauge $a_0 = A_0 = 0$.

$$\Pi_{\alpha\beta}^{\mathrm{CF}}(\boldsymbol{q},\mathrm{i}\omega_n) = -\langle j_\alpha(\boldsymbol{q},\mathrm{i}\omega_n) j_\beta(-\boldsymbol{q},-\mathrm{i}\omega_n)\rangle \tag{5.3.18}$$

is the current-correlation function of the composite fermion. Abberating A_{eff} as $A_{\mathrm{eff}} = 1/2\delta\boldsymbol{a}\hat{\Pi}_{\mathrm{C.S.}}\delta\boldsymbol{a} + 1/2(\delta\boldsymbol{a}+\delta\boldsymbol{A})\hat{\Pi}_{\mathrm{CF}}(\delta\boldsymbol{a}+\delta\boldsymbol{A})$ and integrating with respect to $\delta\boldsymbol{a}$, we obtain the effective action for $\delta\boldsymbol{A}$:

$$A(\{\delta A\}) = \frac{1}{2}\sum_{\boldsymbol{q},\mathrm{i}\omega_n} \delta\boldsymbol{A}(-\boldsymbol{q},-\mathrm{i}\omega_n)\hat{\Pi}(\boldsymbol{q},\mathrm{i}\omega_n)\delta\boldsymbol{A}(\boldsymbol{q},\mathrm{i}\omega_n) . \tag{5.3.19}$$

$\Pi(\boldsymbol{q},\mathrm{i}\omega_n)$ is the 2×2 matrix given by

$$\Pi^{-1}(\boldsymbol{q},\mathrm{i}\omega_n) = \hat{\Pi}_{\mathrm{CF}}^{-1}(\boldsymbol{q},\mathrm{i}\omega_n) + \hat{\Pi}_{\mathrm{C.S.}}^{-1}(\boldsymbol{q},\mathrm{i}\omega_n) . \tag{5.3.20}$$

This corresponds to the composition rule mentioned in (5.2.26). The conductivity tensor of the system is given by

$$\sigma_{\alpha\beta}(\mathrm{i}\omega_n) = \frac{\Pi_{\alpha\beta}(\mathbf{0}, \mathrm{i}\omega_n)}{\mathrm{i}\omega_n}\,. \tag{5.3.21}$$

(Here, we assumed $\Pi(\mathbf{0}, 0) = 0$.) Then, the resistivity tensor of the system is given by

$$\hat{\rho}(\mathrm{i}\omega_n) = \hat{\sigma}^{-1}(\mathrm{i}\omega_n) = \mathrm{i}\omega_n \hat{\Pi}^{-1}(\mathbf{0}, \mathrm{i}\omega_n) \tag{5.3.22}$$

and when the units $\hbar$ and e are recovered in (5.3.20), we obtain

$$\rho_{\alpha\beta}(\mathrm{i}\omega_n) = \rho^{\mathrm{CF}}_{\alpha\beta}(\mathrm{i}\omega_n) + 2\theta\frac{\hbar}{e^2}\varepsilon_{\alpha\beta}\,. \tag{5.3.23}$$

For $\theta = 2\pi n$, when just p Landau levels are filled with composite fermions, we obtain the values

$$\begin{aligned} \rho^{\mathrm{CF}}_{xx} &= \rho^{\mathrm{CF}}_{yy} = 0\,, \\ \rho^{\mathrm{CF}}_{xy} &= -\rho^{\mathrm{CF}}_{yx} = \frac{\hbar}{e^2}\frac{1}{p}\,, \end{aligned} \tag{5.3.24}$$

and therefore when inserting these expressions into (5.3.23), we obtain

$$\begin{aligned} \rho_{xx} &= \rho_{yy} = 0\,, \\ \rho_{xy} &= -\rho_{yx} = \frac{1}{p}\frac{h}{e^2} + 2n\frac{h}{e^2} = \frac{h}{e^2}\left(\frac{1}{p} + 2n\right) = \frac{h}{e^2}\frac{1}{\nu_{n,p}}\,. \end{aligned} \tag{5.3.25}$$

The inverse matrix is given by

$$\begin{aligned} \sigma_{xx} &= \sigma_{yy} = 0\,, \\ \sigma_{xy} &= -\sigma_{yx} = -\frac{e^2}{h}\nu_{n,p}\,. \end{aligned} \tag{5.3.26}$$

The fluctuations of the Chern–Simons gauge field are described by the action (5.3.17) when δA_μ is set to zero, $\delta A_\mu = 0$. Here, we choose the Coulomb gauge $\nabla \cdot \boldsymbol{a} = 0$:

$$A_{\mathrm{gauge}} = \frac{1}{2}\sum_{\boldsymbol{q},\mathrm{i}\omega_n} \begin{bmatrix} \delta a_0 \\ \delta a_1 \end{bmatrix}_{(-\boldsymbol{q},-\mathrm{i}\omega_n)} \begin{bmatrix} \Pi^{\mathrm{CF}}_{00} & \dfrac{\mathrm{i}q}{2\theta} \\ -\dfrac{\mathrm{i}q}{2\theta} & \Pi^{\mathrm{CF}}_{11} - \dfrac{q^2 v(\boldsymbol{q})}{4\theta^2} \end{bmatrix} \begin{bmatrix} \delta a_0 \\ \delta a_1 \end{bmatrix}_{(\boldsymbol{q},\mathrm{i}\omega_n)}\,. \tag{5.3.27}$$

Here, δa_0 is the scalar component, and δa_1 is the transversal component.

The term in the above equation containing $v(\boldsymbol{q})$ expresses the electron-electron interaction

$$\frac{1}{2}\sum_{\boldsymbol{q},\mathrm{i}\omega_n} \delta\rho(-\boldsymbol{q}, -\mathrm{i}\omega_n)v(\boldsymbol{q})\delta\rho(\boldsymbol{q}, \mathrm{i}\omega_n) \tag{5.3.28}$$

through equation (5.3.2), that is $\delta\rho(\boldsymbol{q}, \mathrm{i}\omega_n) = \mathrm{i}q/2\theta\delta a_{\perp}(\boldsymbol{q}, \mathrm{i}\omega_n)$. We assume that Π^{CF} is irreducible with respect to the Coulomb interaction. In the simplest RPA approximation, the free fermion expression Π^{CF} is used. Under the assumption that the fluctuations of the gauge field can be evaluated perturbatively, the picture of the composite fermion is justified. We can assume that this condition is fulfilled when there is a gap in the excitation spectrum of the composite fermions, and the system is a non-compressible liquid.

On the other hand, this is not at all obvious, for example for $\nu = 1/2$, when a large number of low-energy excitations of the composite fermions at the Fermi surface arises. In the limit $\omega \ll v_{\mathrm{F}} \ll \varepsilon_{\mathrm{F}}$, as in (5.2.28), we obtain

$$\begin{aligned} \Pi_{00}^{\mathrm{CF}} &= \frac{m^*}{2\pi} , \\ \Pi_{11}^{\mathrm{CF}} &= -\chi_{\mathrm{CF}} q^2 + \mathrm{i}\omega\gamma_q . \end{aligned} \tag{5.3.29}$$

Here, χ_{CF} is the Landau diamagnetic susceptibility, and γ_q is given by $\gamma_q = 2\bar{\rho}/(m^* q v_{\mathrm{F}})$. Therefore, the propagator of the gauge field $D_{\mu\nu}$ becomes

$$D_{\mu\nu}^{-1}(\boldsymbol{q}, \mathrm{i}\omega_n) = \begin{bmatrix} \dfrac{m^*}{2\pi} & +\dfrac{\mathrm{i}q}{2\theta} \\ -\dfrac{\mathrm{i}q}{2\theta} & \mathrm{i}\gamma_q\omega - \tilde{\chi}(\boldsymbol{q})q^2 \end{bmatrix} , \tag{5.3.30}$$

$$\tilde{\chi}(\boldsymbol{q}) = \chi_{\mathrm{CF}} + \frac{v(\boldsymbol{q})}{2\theta^2} . \tag{5.3.31}$$

Because for the Coulomb interaction, $v(\boldsymbol{q})$ behaves like $v(\boldsymbol{q}) \propto q^{-1}$, the second term on the right-hand side of (5.3.31) becomes dominant for small q, $q^2\tilde{\chi}(\boldsymbol{q}) \propto q$. On the other hand, for a short-range interaction, we obtain $\tilde{\chi}(\boldsymbol{q}) \propto$ const. In this case, the propagator $D_{11}(\boldsymbol{q}, \mathrm{i}\omega_n)$ takes the same form as in the previous section. Therefore, the composite fermion model is similar to the model of spinons presented in the previous section. When the fluctuations of the gauge field are treated perturbatively, singularities like divergence of the self-energy of the fermion occur. This is caused by the fact that $D_{11}(\boldsymbol{q}, \mathrm{i}\omega_n)$ is dominated by the low-frequency, long-wavelength region. It is still an unresolved problem if the composite fermions are a Fermi liquid or not, and whether or not the $\nu = 1/2$ composite fermion model is appropriate for the description. Intensive research is proceeding at present. However, there is no doubt that the concept of the gauge field provides a new and unifying point of view for strongly correlated electronic systems, as for example quantum spin systems, high-T_{c} superconductors and quantum Hall liquids.

Appendix

A. Complex Functions

With x and y being real and i fulfilling $\mathrm{i}^2 = -1$, we define the complex number z by

$$z = x + \mathrm{i}y\,. \tag{A.1}$$

The two-dimensional plane spanned by the coordinates (x, y) is called the complex plane (z-plane). We define the complex conjugate z^* of z by

$$z^* = x - \mathrm{i}y\,. \tag{A.2}$$

Next, we consider a function $f(z)$ of z. Since $f(z)$ also is a complex number, we can express it with real functions $u(z) = u(x, y)$ and $v(z) = v(x, y)$ by

$$f(z) = u(z) + \mathrm{i}v(z)\,. \tag{A.3}$$

$f(z)$ is called 'regular at z_0' when the differential quotient

$$f'(z_0) = \lim_{\Delta z \to 0} \frac{f(z_0 + \Delta z) - f(z)}{\Delta z} \tag{A.4}$$

is finite and continuous.

In general, the limit defined on the right-hand side of (A.4) depends on the direction from which Δz approaches 0. For the case when $f(z)$ is regular, the limit $|\Delta z| \to 0$ is independent of the argument of Δz. This fact leads to a non-trivial equation. The condition that the limit in the direction $\Delta z = \Delta x$ equals the limit in the direction $\Delta z = \mathrm{i}\Delta y$

$$\frac{\partial f(z)}{\partial x} = \frac{1}{\mathrm{i}} \frac{\partial f(z)}{\partial y} \tag{A.5}$$

is equivalent to the Cauchy–Riemann equations

$$\frac{\partial u(x,y)}{\partial x} = \frac{\partial v(x,y)}{\partial y}\,, \qquad \frac{\partial u(x,y)}{\partial y} = -\frac{\partial v(x,y)}{\partial x}\,. \tag{A.6}$$

From (A.6) the Laplace equation

$$\frac{\partial^2 u}{\partial x^2} + \frac{\partial^2 u}{\partial y^2} = 0\,, \qquad \frac{\partial^2 v}{\partial x^2} + \frac{\partial^2 v}{\partial y^2} = 0 \tag{A.7}$$

can be deduced. In a very rough manner, '$f(z)$ is regular at z' can be understood as in the vincinity of z '$f(z)$ can be expressed by z only (without z^*), furthermore, at z_0 no singular behaviour (divergence) arises'.

A.1 Projection from the z-Plane to the w-Plane

$w = f(z)$ is an angle-conserving mapping. In order to see this, we consider z_0 and two neighbouring points z_1 and z_2:

$$\begin{aligned} w_1 - w_0 &= f'(z_0)(z_1 - z_0)\,, \\ w_2 - w_0 &= f'(z_0)(z_2 - z_0)\,, \end{aligned} \tag{A.8}$$

and recall that $f'(z_0)$ in both cases is equal. From (A.8), we obtain

$$\frac{w_2 - w_0}{w_1 - w_0} = \frac{z_2 - z_0}{z_1 - z_0} \tag{A.9}$$

and therefore the angle $< w_2 w_0 w_1$ equals $< z_2 z_0 z_1$.

A.2 Contour Integral of $f(z)$ Around the Path C

We define the integral $\int_C \mathrm{d}z\, f(z)$ in the following manner:

$$\int_C f(z)\,\mathrm{d}z = \lim_{\substack{|z_{i+1}-z_i|\to 0 \\ N\to\infty}} \sum_{i=0}^{N-1} (z_{i+1} - z_i) f(z_i)\,. \tag{A.10}$$

Here, z_0 is the starting point of C, and z_N is the end point of C. For the case when the path C is closed and when $f(z)$ is regular in the area S surrounded by C, we obtain

$$\oint_C f(z)\,\mathrm{d}z = 0 \tag{A.11}$$

(Cauchy's theorem). This theorem can be prooved using Green's integral formula

$$\begin{aligned} \oint_C f(z)\,\mathrm{d}z &= \oint_C (u\,\mathrm{d}x - f\,\mathrm{d}y) + \mathrm{i} \oint_C (u\,\mathrm{d}y + v\,\mathrm{d}x) \\ &= -\iint_S \left(\frac{\partial u}{\partial y} + \frac{\partial v}{\partial x}\right) \mathrm{d}x\,\mathrm{d}y + \mathrm{i} \iint_S \left(\frac{\partial u}{\partial x} - \frac{\partial v}{\partial y}\right) \mathrm{d}x\,\mathrm{d}y \end{aligned} \tag{A.12}$$

and equation (A.6) again.

For the case when $f(z)$ has a pole of order n at $z = z_0$, for $z \to z_0$ with a_{-n} being a coefficient $f(z)$ behaves as

$$f(z) \sim \frac{\alpha_{-n}}{(z-z_0)^n} \,. \tag{A.13}$$

In this case, $f(z)$ can be expanded around z_0 in a Laurant series

$$f(z) = \sum_{l=-n}^{\infty} \alpha_l (z-z_0)^l \,. \tag{A.14}$$

Then, the integration around an infinitesimal closed path C around z_0 leads to

$$\oint_{C_0} f(z)\,\mathrm{d}z = \sum_{l=-n}^{\infty} \alpha_l \oint_{C_0} (z-z_0)^l\,\mathrm{d}z = 2\pi\mathrm{i}\alpha_{-1} \,. \tag{A.15}$$

Here, we used

$$\oint_{C_0} (z-z_0)^l\,\mathrm{d}z = 2\pi\mathrm{i}\delta_{l,-1} \tag{A.16}$$

and defined that C is traced in an anti-clockwise direction. The coefficient a_{-1} is called the residue $\mathrm{Res}\, f(z_0)$ of $f(z)$ at z_0. As a generalization of (A.15), using also (A.14) and (A.16), we obtain

$$\alpha_l = \frac{1}{2\pi\mathrm{i}} \oint_{C_0} \frac{f(z)}{(z-z_0)^{l+1}}\,\mathrm{d}z' \,. \tag{A.17}$$

Integrating $f(z)$ around an arbitrary closed path C, and writing z_i for the singular points of $f(z)$, we obtain by combining Cauchy's theorem with (A.15)

$$\oint_C f(z)\,\mathrm{d}z = 2\pi\mathrm{i} \sum_i \mathrm{Res}\, f(z_i) \,. \tag{A.18}$$

B. The Variational Principle and the Energy–Momentum Tensor

In what follows, we consider the real-time formalism. The action integral A is given by

$$A = \int \mathrm{d}\boldsymbol{r}\,\mathrm{d}t\,\mathcal{L}(\phi, \partial_\mu\phi) = \int \mathrm{d}x\,\mathcal{L}(\phi(x), \partial_\mu\phi(x)) \,, \tag{B.1}$$

with the Lagrangian density $\mathcal{L}(\phi, \partial_\mu\phi)$. We define the four-vector $x_\mu = (\mathrm{i}vt, \boldsymbol{r})$. Then, the distinction between x_μ and x^μ is no longer necessary, and also changing to the imaginary-time formalism is very easy. The field equation is obtained by taking the variation of A

$$\delta A = \int \mathrm{d}x \left(\frac{\partial \mathcal{L}}{\partial \phi} \delta\phi + \frac{\partial \mathcal{L}}{\partial(\partial_\mu \phi)} \partial_\mu (\delta\phi) \right)$$
$$= \int \mathrm{d}x \, \delta\phi \left(\frac{\partial \mathcal{L}}{\partial \phi} - \partial_\mu \frac{\partial \mathcal{L}}{\partial(\partial_\mu \phi)} \right) = 0 \quad \text{(B.2)}$$

leading to

$$\frac{\partial \mathcal{L}}{\partial \phi} - \partial_\mu \frac{\partial \mathcal{L}}{\partial(\partial_\mu \phi)} = 0 \,. \quad \text{(B.3)}$$

We now consider the coordinate transformation $x_\mu \to x_\mu + \delta x_\mu$. Then, the infinitesimal volume element $\mathrm{d}x$ alters to

$$\mathrm{d}x \Rightarrow (1 + \partial_\nu(\delta x_\nu)) \, \mathrm{d}x \,. \quad \text{(B.4)}$$

We use the convention that indices appearing twice are summed over. The Lagrangian density becomes

$$\mathcal{L}(\phi, \partial_\mu \phi) \Rightarrow \mathcal{L}(\phi, \partial_\mu \phi) + \frac{\partial \mathcal{L}}{\partial x_\mu} \delta x_\mu \,. \quad \text{(B.5)}$$

Considering $\partial \mathcal{L} / \partial x_\mu$ in more detail, we obtain

$$\frac{\partial \mathcal{L}}{\partial x_\mu} = \frac{\partial \mathcal{L}}{\partial \phi} \partial_\mu \phi + \frac{\partial \mathcal{L}}{\partial(\partial_\nu \phi)} \partial_\mu \partial_\nu \phi \,. \quad \text{(B.6)}$$

Using the equation of motion (B.3), we obtain

$$\frac{\partial \mathcal{L}}{\partial x_\mu} = \partial_\nu \left(\frac{\partial \mathcal{L}}{\partial(\partial_\nu \phi)} \right) \cdot \partial_\mu \phi + \frac{\partial \mathcal{L}}{\partial(\partial_\nu \phi)} \partial_\nu (\partial_\mu \phi)$$
$$= \partial_\nu \left[\frac{\partial \mathcal{L}}{\partial(\partial_\nu \phi)} \partial_\mu \phi \right] , \quad \text{(B.7)}$$

and finally the variation δA of the action becomes

$$\delta A = \int \mathrm{d}x \left(\partial_\nu(\delta x_\nu) \mathcal{L} + (\delta x_\mu) \partial_\nu \left[\frac{\partial \mathcal{L}}{\partial(\partial_\nu \phi)} \partial_\mu \phi \right] \right)$$
$$= \int \mathrm{d}x \left(\delta_{\mu\nu} \mathcal{L} - \frac{\partial \mathcal{L}}{\partial(\partial_\nu \phi)} \partial_\mu \phi \right) \partial_\nu (\delta x_\mu) \,. \quad \text{(B.8)}$$

Writing δA as

$$\delta A = -\frac{1}{2\pi} \int \mathrm{d}x \, T_{\mu\nu} \partial_\mu (\delta x_\nu) \quad \text{(B.9)}$$

we obtain

$$T_{\mu\nu} = -2\pi \left(\delta_{\mu\nu} \mathcal{L} - \frac{\partial \mathcal{L}}{\partial(\partial_\mu \phi)} \partial_\nu \phi \right) . \quad \text{(B.10)}$$

Here,

$$T_{00} = -2\pi \left(\mathcal{L} - \frac{\partial \mathcal{L}}{\partial(\partial_0 \phi)} \partial_0 \phi \right) \tag{B.11}$$

is 2π times the Hamiltonian density. Therefore, we obtain

$$H = \int \frac{\mathrm{d}\boldsymbol{r}}{2\pi} T_{00}(x) \, . \tag{B.12}$$

In special relativity, the momentum four-vector P_μ consists of the usual momentum three-vector as space coordinate, and in the time coordinate (the zeroth component) of the energy $\times(\mathrm{i}/v)$. Introducing also this coefficient of the zeroth component, we obtain

$$P_\mu = -\mathrm{i}v \int \frac{\mathrm{d}\boldsymbol{r}}{2\pi} T_{\mu 0}(x) \, . \tag{B.13}$$

Requiring invariance of the action under the coordinate transformation $\delta x_\mu =$ constant, the right-hand side of (B.7) must be zero, which translates via (B.10) into

$$\partial_\nu T_{\mu\nu} = 0 \, . \tag{B.14}$$

Because, in general, $\partial_\nu A_\nu = 0$ signifies

$$\operatorname{div} \boldsymbol{A} + \frac{1}{\mathrm{i}v} \frac{\partial A_0}{\partial t} = 0 \tag{B.15}$$

the space integral over A_0 is a conserved quantity. Therefore, (B.13) is a conserved quantity. Above, we mentioned the case of real-time formalism. For the case of imaginary-time formalism, that is important in this book, notice that many signs do change.

Literature

For this book, the knowledge of the volume

[G.1] N. Nagaosa: Quantum Field Theory in Condensed Matter Physics (Springer, Berlin, Heidelberg, 1999) is presumed. Some general books about quantum field theory that have not been mentioned therin are:
[G.2] M.E. Peskin and D.V. Schroeder: *An Introduction to Quantum Field Theory* (Addison-Wesley, 1995)
[G.3] K. Huang: *Quarks, Leptons and Gauge Fields* (World Scientific, 1992)
[G.4] T.P. Cheng and L.F. Li: *Gauge Field Theory of Elementary Particle Physics* (Cambridge Univ. Press, 1995)
[G.5] K. Yoshida: Theory of Magnetism (Springer Series in Solid-State Sciences, 122, 1996)
[G.6] A.M. Tsvelik: Quantum Field Theory in Condensed Matter Physics (Cambridge Univ. Press, 1995)

Chapter 1

About the Bethe ansatz:

[1] H.B. Thacker: Rev. Mod. Phys. 53 (1981) 253
[2] B.S. Shastry et al. eds.: *Exactly Solvable Problems in Condensed Matter and Relativistic Problems*, Lecture Notes in Physics 242 (Springer, Berlin, Heidelberg, 1985)

Chapter 2

About Bosonization:

[G.7] A.O. Gogolin, A.A. Nersesyan, and A.M. Tsvelik: Bosonization and Strongly Correlated Systems (Cambridge Univ. Press, 1998)
[3] J. Solyom: Adv. Phys. 28 (1979) 201
[4] V.J. Emery: in *Highly Conducting One-Dimensional Solids*, JT. Devreese et al. eds. (Plenum Press, 1979) p. 247

[5] H. Fukuyama and H. Takayama: in *Electronic Properties of Inorganic Quasi-One-Dimensional Compounds*, P. Monceau ed. (D. Reidel, 1985) p. 41
About the analysis of the spin chain using the non-linear sigma model:
[6] F.D. Haldane: Phys. Rev. Lett. 50 (1983) 1153
The analysis of the same model in two dimensions is done in:
[7] S. Chakravarty, B.I. Halperin, and D.R. Nelson: Phys. Rev. Lett. 60 (1988) 1057

Chapter 3

The spin/charge separation in one dimension is analysed in [3], [4], [5]. The SCR theory is described and compared with the experiments in

[8] T. Moriya: *Spin Fluctuations in Itinerant Electron Magnetism*. Solid-State Science 56 (Springer, Berlin, Heidelberg, 1985)
The quantum renormalization group is discussed in
[9] J.A. Hertz: Phys. Rev. B14 (1976) 1165
[10] A.J. Millis: Phys. Rev. B48 (1993) 7183
A review on metal–insulator transition is
[11] M. Imada, A. Fujimori, and Y. Tokura: Rev. Mod. Phys. 70 (1998) 1039

Chapter 4

[12] P. Coleman: Phys. Rev. B35 (1987) 5072
[13] D.L. Cox and A. Ruckenstein: Phys. Rev. Lett. 71 (1993) 1613
A good textbook about the Kondo problem is
[14] A.C. Hewson: *The Kondo Problem to Heavy Fermions* (Cambridge Univ. Press, 1993)
The original paper of NCA is
[15] Y. Kuramoto, Z. Phys. B53 (1983) 37.
A comprehensive review about the dynamical mean field theory is provided by
[16] A. Georges, G. Kotliar, W. Krauth, and M.J. Rosenberg: Rev. Mod. Phys. 68 (1996) 13

Chapter 5

Concerning the content of Sect. 5.1:

[17] I. Affleck: in *Strings, Fields and Critical Phenomena*, E. Brezin and J. Zinn-Justin eds. (North-Holland, 1990) p. 565

Concerning the content of Sect. 5.2:

RVB theories of high-T_c superconductivity originate from

[18] P.W. Anderson: Science 235 (1987) 1196

See also

[19] P.W. Anderson: *The Theory of Superconductivity in the High T_c Cuprates* (Princeton, 1997)

[20] Y. Suzumura, Y. Hasegawa, and H. Fukuyama, J. Phys. Soc. Jpn. 57 (1988) 2768

[21] P.A. Lee and N. Nagaosa: Phys. Rev. B46 (1992) 5621

An important paper about the $\nu = 1/2$ Landau level discussed in Sect. 5.3:

[22] B.I. Halperin, P.A. Lee, and N. Read: Phys. Rev. B47 (1993) 7312

Index

Printing (Computer to Film): Saladruck, Berlin
Binding: Lüderitz & Bauer, Berlin

Printed in the USA
CPSIA information can be obtained
at www.ICGtesting.com
LVHW011524030923
757103LV00005B/71